Hydrocarbon Chemistry

Hydrocarbon Chemistry

Editor

Sarika Srivastav

Hydrocarbon Chemistry
Edited by **Sarika Srivastav**

Printed in 2017

ISBN: 978-1-68117-403-7

Library of Congress Control Number: 2015941599

© 2016 by
SCITUS Academics LLC,
616, Corporate Way, Suite 2, 4766,
Valley Cottage, NY 10989

www.scitusacademics.com

This book contains information obtained from highly regarded resources. Copyright for individual articles remains with the authors as indicated. All chapters are distributed under the terms of the Creative Commons Attribution License, which permits unrestricted use, distribution, and reproduction in any medium, provided the original author and source are credited.

Notice

Reasonable efforts have been made to publish reliable data and views articulated in the chapters are those of the individual contributors, and not necessarily those of the editors or publishers. Editors or publishers are not responsible for the accuracy of the information in the published chapters or consequences of their use. The publisher believes no responsibility for any damage or grievance to the persons or property arising out of the use of any materials, instructions, methods or thoughts in the book. The editors and the publisher have attempted to trace the copyright holders of all material reproduced in this publication and apologize to copyright holders if permission has not been obtained. If any copyright holder has not been acknowledged, please write to us so we may rectify.

Contents

Preface vii

Chapter 1   Influence of Decomposition Time and H_2 Pressure on Properties of Unsupported Ammonium Tetrathiomolybdate-derived Mos2 Catalysts 1

Jamie Whelan, Ionut Banu, Gisha E Luckachan, Nicoleta Doriana Banu, Samuel Stephen, Anjana Tharalekshmy, Saleh Al Hashimi, Radu V Vladea, Marios S Katsiotis, and Saeed M Alhassan

Chapter 2   Effects of Multiwalled Carbon Nanotubes and Triclocarban on Several Eukaryotic Cell Lines: Elucidating Cytotoxicity, Endocrine Disruption, and Reactive Oxygen Species Generation 25

Anne Simon, Sibylle X Maletz, Henner Hollert, Andreas Schäffer, and Hanna M Maes

Chapter 3   Optimal Parameters for in Vitro Development of the Fungus Hydrocarbonoclastic Penicillium sp. 69

Marcia Eugenia Ojeda-Morales, Miguel Ángel Hernández-Rivera, José Gabriel Martínez-Vázquez, Yolanda Córdova-Bautista, and Yuridia Evelin Hernández-Cardeño

Chapter 4   An Application of the Taguchi Method (Robust Design) to Environmental Engineering: Evaluating Advanced Oxidative Processes in Polyester-Resin Wastewater Treatment 99

Messias Borges Silva, Livia Melo Carneiro, João Paulo Alves Silva, Ivy dos Santos Oliveira, Hélcio José Izário Filho, and Carlos Roberto de Oliveira Almeida

Chapter 5   Effect of Two Liquid Phases on the Separation Efficiency of Distillation Columns 121

Gardênia Marinho Cordeiro, Stephanie Rolim Dantas, Luís Gonzaga Sales Vasconcelos, and Romildo Pereira Brito

Chapter 6	Catalyst Deactivation and Engineering Control for Steam Reforming of Higher Hydrocarbons in a Novel Membrane Reformer 141
	Zhongxiang Chen, Yibin Yan, and Said S.E.H. Elnashaie
Chapter 7	In Silico Bioremediation of Polycyclic Aromatic Hydrocarbon: A Frontier in Environmental Chemistry 179
	Vito Librando and Matteo Pappalardo
Chapter 8	Analysis of Gas Phase Compounds in Chemical Vapor Deposition of Carbon from Light Hydrocarbons 211
	Koyo Norinaga, Olaf Deutschmann, and Klaus J. Hüttinger
Chapter 9	Adsorption of Mixed Polycyclic Aromatic Hydrocarbons in Surfactant Solutions by Activated Carbon 243
	Jianfei Liu, Jiajun Chen, Lin Jiang, and Xue Yin

Citations 273

Index 277

Preface

Hydrocarbons and their transformations play major roles in chemistry as raw materials and sources of energy. Diminishing petroleum supplies, regulatory problems, and environmental concerns constantly challenge chemists to rethink and redesign the industrial applications of hydrocarbons. Hydrocarbons are ubiquitous ingredients of the chemical composition of the troposphere. While present as trace components, they make a major contribution toward the production of ozone and other oxidants such as peroxyacetyl nitrate (PAN) and hydrogen peroxide (H202). Hydrocarbon Chemistry begins by discussing the general aspects of hydrocarbons, the separation of hydrocarbons from natural sources, and the synthesis from C1 precursors with recent developments for possible future applications. Carbon and Hydrogen: the two basic building units can be combined in a million different ways to give a plethora of fascinating organic compounds.

Editor

Chapter 1

Influence of Decomposition Time and H_2 Pressure on Properties of Unsupported Ammonium Tetrathiomolybdate-derived MoS_2 Catalysts

Jamie Whelan[1,2], Ionut Banu[3], Gisha E Luckachan[1], Nicoleta Doriana Banu[1,4], Samuel Stephen[1], Anjana Tharalekshmy[1], Saleh Al Hashimi[1], Radu V Vladea[1], Marios S Katsiotis[1], and Saeed M Alhassan[1]

[1]Department of Chemical Engineering, The Petroleum Institute, Abu Dhabi, United Arab Emirates
[2]Department of Chemistry, New York University Abu Dhabi, Abu Dhabi, UAE.
[3]Department of Chemical and Biochemical Engineering, University

Politehnica of Bucharest, 313 Spl. Independentei, sector 6, Bucharest, 060042, Romania

[4]Center for Organic Chemistry "C.D. Nenitzescu", Bucharest, 060023, Romania

ABSTRACT

Background

Molybdenum sulfide (MoS_2) catalysts to be used for hydrodesulfurization (HDS) processes were prepared via the reductive thermal decomposition of ammonium tetrathiomolybdate at fixed temperature (653 K) by varying decomposition times and H_2 pressures. Both parameters were found to strongly influence textural and catalytic properties of the resulting MoS_2 catalysts.

Methods

Nitrogen sorption, FT-IR, and XRD analyses revealed the effect of varying decomposition times (3 to 7 h) and H_2 pressure (20 to 1,000 psig) on the morphology and structure of the catalysts. Dibenzothiophene (DBT) was used to assess catalytic efficiency for HDS reactions.

Results

The influence of time on specific surface was minimal at low pressures but increased at higher decomposition pressures. Vibrational energies of Mo-S bonds in FT-IR indicate that MoS_2 catalysts prepared at higher pressures exhibit weaker Mo-S bonds. Analysis of XRD patterns point towards an increase in stacking and crystallite size with increasing pressure; interlayer rotation about both the a- and c-axes of the stacks was also observed. Catalytic testing results show that conversion increases at higher values of decomposition time and pressure. Partially hydrogenated products were also observed at higher pressures, and the ratio of partially to fully hydrogenated DBT was calculated as an additional measure of catalytic efficiency.

Conclusions

Decomposition time and H_2 pressure during ammonium tetrathiomolybdate (ATM) thermal decomposition have a significant impact on the morphological and catalytic properties of the derived MoS_2 catalysts. Samples prepared for 5 h at 1,000 psig exhibited the highest conversion of DBT and the lowest ratio of partially to fully hydrogenated products.

BACKGROUND

With environmental concerns continually on the rise, greater demand for fuels with low-sulfur content has increased the focus on hydrotreating (HDT) catalysts (Topsøe et al. [1996]). Molybdenum sulfide (MoS_2)-based catalysts, varyingly promoted with cobalt or nickel, are one of the most common metal sulfides used in HDT, with a strong emphasis on hydrodesulfurization (HDS) reactions (Brunet et al. [2005]; Egorova and Prins [2006]; Álvarez et al. [2008]; Breysse et al. [2008]; Chianelli et al. [2009]; Klimov et al. [2010]). There are several common methods to synthesize this catalyst such as the direct sulfidation of molybdenum oxide, hydrothermal and sonochemical synthesis, and the relatively simple procedure of thermally decomposing ammonium tetrathiomolybdate (ATM) (Camacho-Bragado et al. [2005]; Devers et al. [2002]; Mdleleni et al.[1998]; Afanasiev [2008]; Polyakov et al. [2008]).

Thermal decomposition of ATM is generally a solid-state reaction and, like similar reactions, can be influenced by a number of different experimental parameters such as temperature, gaseous environment, time, and pressure. The proposed reaction mechanism is a two-step process, as can be shown below as Reactions 1 and 2 (Walton et al. [1998]):

$$(NH_4)_2MoS_4 + H_2 \rightarrow MoS_3 + H_2S + 2NH_3 \tag{1}$$

$$MoS_3 + H_2 \rightarrow MoS_2 + H_2S \tag{2}$$

While several studies of note have already examined the effect of pre-treatment of ATM and the correlation between MoS_2 HDS activity by varying gaseous environment (e.g., H_2S/H_2) and Mo precursors as well as reducing and sulfiding conditions, little work has focused on the influence of decomposition pressure (Liang et al. [1986]; Zhang and Vasudevan [1995]; Alonso et al. [1998]; Afanasiev [2010]). As a general rule, pressure increases are known to decrease the rate of thermal decomposition; however, other influences have been observed. In their study of nitramine compounds, Piermarini et al. found a change in decomposition mechanism above a certain pressure (Piermarini et al. [1987]). In addition, Criado et al. studied the decomposition of $CaCO_3$ and observed a shift towards higher decomposition temperatures with increasing CO_2 pressure (Criado et al. [1995]).

Though some examples in literature report pressure-induced effects on MoS_2 properties such as electrical conductivity and critical current density, little is known regarding the pressure effects on decomposition of ATM and HDS activity of unsupported MoS_2 catalyst (Sánchez et al. [2006]; Alekseevskii et al. [1977]). While it could be argued that the impact of decomposition may be small relative to other 'stronger' experimental parameters, clearly, the studies on decomposition of nitramine compounds and $CaCO_3$ show it to be a potentially fruitful exercise. Thus, the present study focuses on pressure and time effects on reductive thermal decomposition of ATM and the catalytic activity of the resulting MoS_2. It is shown that by varying the decomposition parameters, resulting structural and chemical differences prove insightful towards designing an economical catalyst for the effective removal of sulfur. The degree of influence these experimental parameters have on the morphology and activity of MoS_2 catalysts is described below.

METHODS

Reagents and Solutions

The following chemicals were used as purchased: ATM (Aldrich, Dorset UK, 99.97%), dibenzothiophene (DBT) (Aldrich, 98%), hexadecane (Aldrich, 99%), hydrogen gas (Air Products, 99.992%), and nitrogen gas (Air Products, Allentown, PA, USA; 99.995%).

Catalyst Preparation

Ammonium tetrathiomolybdate was weighted on an aluminum boat and placed into a 300-cm^3 batch reactor at room temperature. The reactor was sealed, flushed with N_2 and H_2, and then heated to 623 K (10 K min^{-1}). Upon reaching the reaction temperature, H_2 was injected into the reactor at the required pressure, taken as time zero. After the chosen heating duration was completed, heating would be stopped and the reactor would be allowed to cool naturally while maintaining a reductive (H_2) atmosphere. Upon reaching room temperature, the reactor was degassed, flushed with N_2, opened, and the resulting product (MoS_2) was removed and weighted. Immediately afterwards, the specimen was ground using a pestle and mortar and used for characterization and HDS testing. MoS_2 specimens prepared by this method were named as follows: number (time in hours)-number (hydrogen pressure, psig); decomposition times were 3, 5, and 7 h and H_2 pressures were 20 (0.138 MPa), 500 (3.447 MPa), and 1,000 psig (6.895 MPa). For example, 3-1,000 implies MoS_2 prepared from ATM thermally decomposed at 623 K for 3 h at 1,000 psig H_2.

Characterization

Textural characterization was carried out on all catalysts with N_2 sorption at 77 K with a Quantachrome Autosorb-1 (Quantachrome Instruments, Boynton Beach, FL, USA). Prior to analysis, each sample was degassed under vacuum at 573 K for 2 h; the BET surface area (S_{BET}) was determined from the resulting isotherm. FT-IR spectra were obtained using diffuse reflectance infrared Fourier transform (DRIFT) spectroscopy on a Bruker Vertex 70 (Bruker AXS, Inc., Madison, WI, USA). The dark-colored catalysts were mixed with KBr at a ratio MoS_2:KBr = 1:100 to improve infrared transmission; spectra were collected in the 4,000 to 350-cm^{-1} range. Samples for XRD were mounted on a zero background holder, and spectra were collected with a X'Pert PRO Panalytical Powder diffractometer (PANalytical, Almelo, The Netherlands), using Cu K radiation (45 kV and 40 mA) in the 2-theta range 10° to 80° with a step time of 0.01 s. Temperature-programmed reduction (TPR) analyses were carried out under continuous H_2 flow (5% H_2 in He) in a quartz cell at 10 K min^{-1} using a Quantachrome ChemBET 3000 (Quantachrome Instruments, Boynton Beach, FL, USA) coupled with a Hiden Analytical HPR 20 QIC mass spectrometry detector (Hiden Analytical, Warrington, UK). It should be mentioned that in no case was any specimen exposed to air for more than 2 min, thus avoiding oxidation of the sulfide catalysts.

Activity Measurements

Freshly synthesized and ground MoS_2 (0.050 g, 3.1×10^{-4} mol) was added to a 100-cm^3 batch reactor. Feedstock used was 1% DBT in hexadecane (25 cm^3). The reactor was sealed, flushed with N_2, and then heated to 573 K under stirring (1,000 rpm). Upon reaching the reaction temperature, H_2 (500 psig) was injected into the reactor (time zero). Heating and stirring would be stopped after 3 h, and the reactor would be left to cool naturally, while retaining a reductive (H_2) atmosphere. Upon reaching room temperature, the reactor was degassed and a sample of the reaction mixture was removed, centrifuged, and the supernatant diluted and analyzed on a gas chromatograph coupled with a sulfur chemiluminescence detector (GC-SCD - Agilent 6C 6980 and SCD 335 (Agilent Technologies, Inc., Santa Clara, CA, USA)). The column was 100% dimethylpolysiloxane, 30 m × 0.32 mm × 1 μm, with a maximum temperature of 598 K. In this study, conversion of DBT was calculated based on the decrease in DBT signal (and increase in partially hydrogenated DBT (HYD)) from the GC-SCD compared to the initial concentration following calibration.

RESULTS AND DISCUSSION

TPR-MS experiments (Figure 1) show two peaks for the reductive thermal decomposition of ATM in H_2, supporting the proposed ATM decomposition mechanism. Thus, the peak at 475 K is assigned to Reaction (1), the release of H_2S and NH_3 (m/z peaks followed were 16, 17, 32, 33, and 34 amu; for display purposes, only molecular ion peaks are shown), while the peak at 613 K is assigned to Reaction (2), i.e., the release of the second molecule of H_2S.

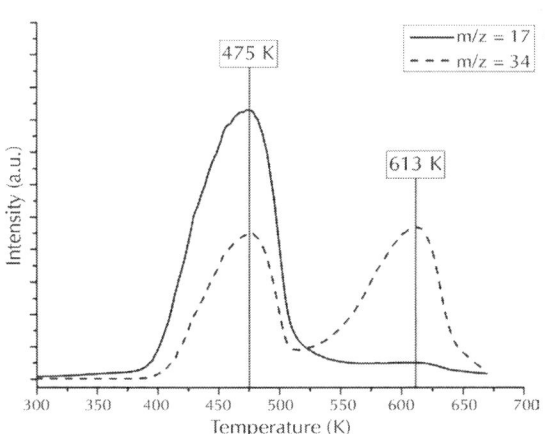

Figure 1: TPR-MS of reductive thermal decomposition of ATM. TPR-MS of reductive thermal decomposition of ATM under H_2 (5% H_2/He) showing evolution of NH_3 (m/z=17) and H_2S (m/z=34).

According to the reaction mechanism, for every mole of ATM that decomposes, four moles of gaseous products are produced ($2NH_3$ and $2H_2S$). Based on the starting concentration of ATM, it was calculated that there was a pressure increase of approximately 15 psig above the H_2 injected upon reaching decomposition temperature; no attempt was made to correct for this during catalyst preparation. In general, throughout this paper, the H_2 decomposition pressures applied are termed as low (20 psig), medium (500 psig), and high (1,000 psig), and it is only against the pressure trend that meaningful conclusions are obtained, not the actual value itself. It should be noted that the actual pressure 'felt' by the ATM is not exactly the same pressure as mentioned above; variation of pressure around the decomposition reaction zone lead to different values than the set point of the external pressure regulator.

Textural Characterization

Textural characterization of the catalysts shows a change in S_{BET} (Figure 2) and total pore volume (TPV) (Figure 3) as a function of decomposition time and pressure, respectively. In general, catalysts prepared at low pressure (20 psig) and low decomposition time (<5 h) exhibited little change in morphology; however, with an increase in pressure, the influence of heating time also increased. Within each time series, increasing pressure resulted in increasing S_{BET} and TPV with the largest changes observed for the 5-h series.

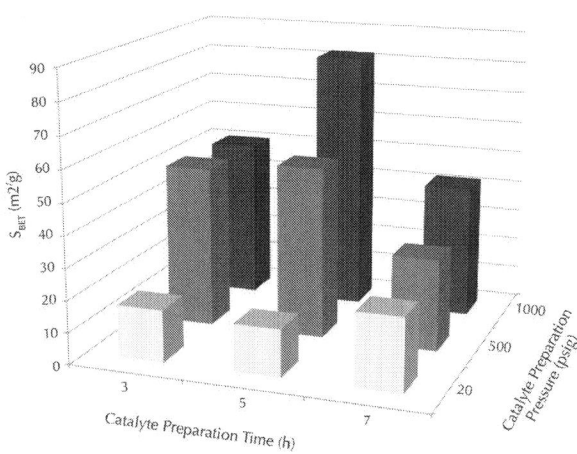

Figure 2: Change in specific surface area (S_{BET}) vs. decomposition time and H_2 pressure.

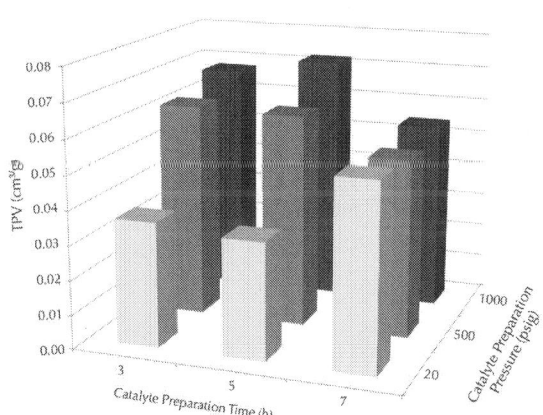

Figure 3: Change in total pore volume (TPV) vs. decomposition time and H2 pressure.

XRD Characterization

All samples display the typical diffraction profile for poorly crystalline MoS_2, namely the diffraction peaks of (002), (100), (103),

(105), and (110). For the sake of article space, only spectra from specimens prepared at 1,000 psig are shown in Figure 4; all spectra can be found in the Additional file 1. Using the Debye-Scherrer equation applied to the broadening of the (110) diffraction peak, it was found that over time there was an increase in crystalline order along the basal direction, likewise with increasing pressure as shown in Table 1 (Daage and Chianelli [1994]). Calculations of d-spacing yield no discernible pattern across the samples (average of 6.30 Å). Using previously reported intensity ratio analysis (e.g., $I_{110}:I_{002}$), it was found that stacking was affected by a change in pressure and with time; using the Debye-Scherrer equation applied to the broadening of the (002) diffraction peak (Table 1), it was found that with increasing decomposition time and pressure the influence on stacking height decreased. However, of note is that while for 3 and 5 h increasing decomposition pressure increased stacking height, for the longest time in this study (7 h), stacking height was found to decrease as decomposition pressure increased. The exact role of stacking height on the catalytic activity of MoS_2 is unclear with support both for and against its influence (Daage and Chianelli [1994]; Afanasiev [2010]).

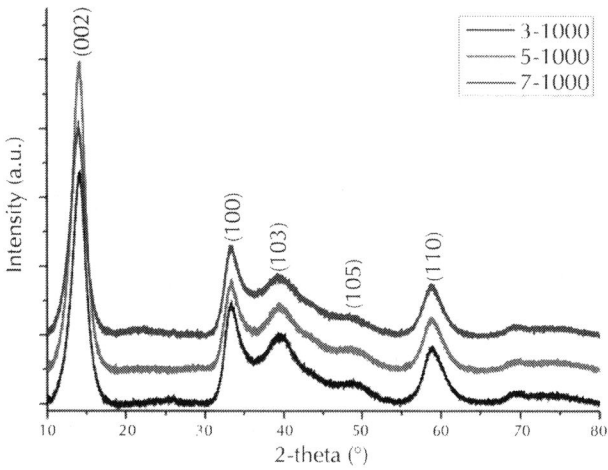

Figure 4: XRD patterns for MoS_2 catalysts prepared by thermal decomposition of ATM at 1,000 psig for varying durations.

Table 1: Crystalline order along the basal direction and apparent stacking heights in the c-axis direction

Pressure (psig)	20			500			1,000		
Time (h)	(110)	(002)	SA (m²/g)	(110)	(002)	SA (m²/g)	(110)	(002)	SA (m²/g)
3	51	23	350	61	28	290	63	30	270
5	54	25	320	63	30	270	61	31	270
7	64	32	260	65	31	260	62	29	280

Crystalline order along the basal direction (Å) and apparent stacking heights in the c-axis direction (Å) calculated using the Debye-Scherrer equation to the broadening of the (110) and (002) diffraction peaks, respectively, for MoS2 catalysts. Surface Area (SA) was calculated in m²/g using the formula mentioned in the work by Iwata et al ([2001]).

Whelan et al.

Whelan et al. Journal of Analytical Science and Technology 2015 6:8 doi:10.1186/s40543-014-0043-0

FT-IR Catalyst Characterization

Five distinct peaks were observed in the FT-IR spectra of the samples in the region below ca. 700 cm^{-1}: 655, 594, 462, 385, and 378 cm^{-1}. All samples were found to contain the first three peaks but varied in the latter two, containing 378 and/or 385 cm^{-1}. The FTIR spectra for MoS$_2$ catalysts prepared for 3-h decomposition time and 1,000 psig are shown in Figures 5 and 6, respectively (normalized against the peak at 378 cm^{-1} for ease of illustration). The peaks at 378 and 385 cm^{-1} have been assigned to Mo-S stretching vibrations along the basal plane (Berhault et al. [2002]). With increasing heating time and pressure, the peak(s) at 378 and/or 385 cm^{-1} shifted towards a single peak at 378 cm^{-1}. The peak at 462 cm^{-1} is generally assigned to the Mo-S vibration perpendicular to the basal plane, i.e., the bridging Mo-S bond (Fedin et al. [1989]). The presence of intermediate MoS_3 (with a peak at 385 cm^{-1}) is discounted due to the lack of corroborating peaks such as a peak at 525 cm^{-1} (Weber et al. [1995]). The peaks at 655 and 594 cm^{-1} can be attributed to sulfur-containing peaks, with that at 655 cm^{-1} assigned to S-H vibrations (from Mo-SH) and that at 594 cm^{-1} attributed to S-S bonding strongly coupled to Mo-S (Müller et al. [1982]).

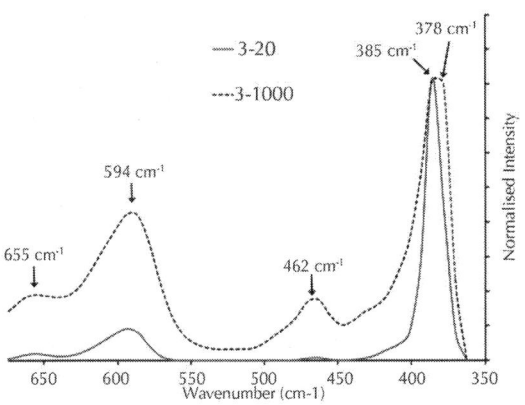

Figure 5: FT-IR spectra for specimens 3-20 and 3-1,000. FT-IR spectra for specimens 3-20 and 3-1,000 showing variation of Mo-S peak vibrations (378 and/or 385 cm^{-1}) and S-S peak (592 cm^{-1}); spectra normalized against peak at 378 cm^{-1}.

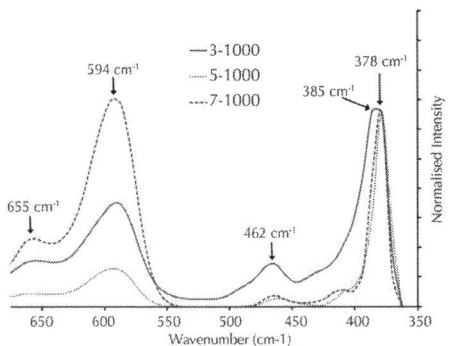

Figure 6: FT-IR spectra for 3-1,000, 5-1,000 and 7-1,000. FT-IR spectra for 3-1,000, 5-1,000, and 7-1,000 showing variation of Mo-S peak vibra-

tions (378 and/or 385 cm⁻¹) and S-S peak (592 cm⁻¹); spectra normalized against peak at 378 cm⁻¹.

Catalyst Activity

HDS activity of each MoS_2 catalyst was tested, and the %DBT conversion of each catalyst is given in Table 2. HDS reaction conditions were set such that catalytic activity was kept below 20%, primarily to remove uncertainties associated with produced H_2S, which could limit the forward reaction and could increase the pressure in the batch reactor. A trend was observed of a general increase in HDS catalytic activity of MoS_2 prepared for longer decomposition times (i.e., across the row), with the exception of 7-1,000, as well as at higher H_2 pressure (i.e., down the column), excluding 7-500 and 7-1,000. The amount of partial HYD is also given below (values in parentheses), and it can be seen that these percentages change depending upon the MoS_2 preparation conditions. As an example, the chromatogram of reaction product for 5-500 showed a reduction of DBT (retention time = 20.4 min) concentration of 12.0% and an increase in HYD (retention time = 20.2 min) product of 2.8%; thus, direct desulfurization (DDS) accounts for 9.2% of DBT removal.

Table 2: Catalytic activity of MoS_2 samples

Pressure (psig)	20	500	1,000
Time (h)			
3	1.7 (0)	5.8 (1.4)	12.0 (1.7)
5	5.8 (0)	12.0 (2.8)	15.0 (2.0)
7	7.7 (0)	10.0 (1.2)	5.2 (1.3)

Catalytic activity of MoS_2 samples defined as % conversion of DBT in hexadecane at 573 K, 500 psig (H_2), for 3 h. Values in parentheses represent the % of total conversion which is attributable to hydrogenated DBT (HYD).

Whelan et al.

Whelan et al. Journal of Analytical Science and Technology 2015 6:8 doi:10.1186/s40543-014-0043-0

Diffusion rates of gaseous products from thermal decomposition of solids are known to decrease with an increase in pressure (Oyumi and Brill [1987]). In the case of ATM decomposing, gas is released from each step (see Equations 1 and 2) leading to an increase in pressure which could slow down diffusion of gaseous by-products. This is expected to have an impact on the resulting MoS_2 formed.

Analysis of Results

A general increase in S_{BET} with an increase in H_2 pressure (Figure 2) was observed across all decomposition times, though to varying degrees. The influence of decomposition time on S_{BET} at low pressure was minimal, though as decomposition pressure increased, the influence of decomposition time became more pronounced. Based on the experimental parameters studied herein, it appears that 5 h was the optimal time and 1,000 psig the optimal pressure to produce a catalyst with the largest specific surface area. Increasing H_2 pressure was also found to produce larger pore volumes in the catalyst (Figure 3) leading to the conclusion that textural properties of the catalyst can be controlled by optimizing ATM decomposition time and pressure.

While increasing H_2 decomposition pressure increased S_{BET} within each decomposition time series, S_{BET} did not increase within each decomposition pressure series, particularly from 5 to 7 h. Intuitively, it is not unreasonable to expect an increase in pressure to result in a greater propensity for pore collapse (and subsequent decrease in surface area) which we see in 7-500 and 7-1,000, when compared against 5-500 and 5-1,000; however, the fact that pore collapse only happens at longer decomposition times again highlights the importance of the combined and sometimes conflicting roles of time and pressure and the necessity of varying ATM decomposition parameters to ensure greater control of the resulting MoS_2. A similar trend was observed for TPV of MoS_2

samples (Figure 3) in that the greatest increase in TPV was for the 5-h series and the lowest was for the 7-h series and likewise a decrease in TPV for the high-pressure 7 h series compared to that for the 5-h series (e.g., compare 5-500 with 7-500 and 5-1,000 with 7-1,000).

The change in stacking height with a change in pressure is not a novel concept having been noted before on supported hydrotreating catalysts, with HDS process conditions determined as the main cause of destacking (De la Rosa et al. [2004]). Hydrothermal synthesis of MoS_2 in an autoclave was also found to result in destacking, and in that study, the presence of hydrocarbons was presented as a possible contributor (Peng et al. [2001]) while, in a separate study, Chianelli et al. note that possible intercalation of H_2 at high pressures could result in MoS_2 layers simply sliding apart over time (Chianelli et al. [2006]). However, all of these analyses were conducted on catalysts under HDS conditions whereas, in this case, stacking is not measured as a function of reaction conditions but as a function of thermal decomposition parameters on the original ATM starting material.

For any crystalline solid (in this case poorly crystalline), it can be expected that with an increase in stacking there should be a corresponding increase in S_{BET}; however, the degree to which S_{BET} increases would not be by the same factor as the increase in stacking, simply on the basis that it is only the exposed edge surface which will add to S_{BET} whereas the basal planes are 'covered' by the extra stacks (this of course infers perfect alignment of stacks). The use of S_{BET}, however, assumes that there is no agglomeration. Iwata et al. have observed vastly different S_{BET} from the theoretical values of surface area obtained from utilizing data from XRD (Iwata et al. [2001]); applying their formula, a similar pattern is evident in our samples, i.e., agglomeration is occurring (compare Figure 2 and SA values in Table 1). While S_{BET} increases with an increase in decomposition pressure, it can be seen from Table 1 that the XRD-determined surface area decreases, appearing to level off close to 270 m^2 g^{-1}. Thus, the pattern emerges within each time series such that the extent of agglomeration is decreasing with an increase in decomposition pressure. Agglomeration is assumed to be prevalent in the solid state, i.e., during N_2 sorption, though less so during activity measurements due to the presence of solvent and high temperatures.

Previously reported modelling studies of poorly crystalline MoS_2 have shown that interlayer rotation about the a-axis results in little change for the 110 peak but in a shift of the 002 peak towards lower angles (Liang et al. [1986]). The same authors model interlayer rotation about the c-axis, with changes observed in the 100-103-105 region, with only a small effect on the 002 and 110 peaks. XRD patterns of our samples display slight variations in the angle of 002, as well changes in $I_{103}:I_{002}$ and $I_{105}:I_{002}$ indicating interlayer rotation about the a- and c-axes, respectively, though by how much exactly is unknown.

The presence of two distinct peaks for Mo-S bonds implies a variation in strength and/or length; based on FT-IR vibrational energies, we assign 385 cm^{-1} to the stronger (i.e., shorter) Mo-S bond, with the weaker (longer) Mo-S bond at 378 cm^{-1}. Mo-S vibrations are usually assigned to the region ca. 380 cm^{-1}, with no literature evidence on FT-IR of MoS_2 found distinguishing between these peaks. In a related study, however, laser Raman studies of alumina-supported MoS_2 by Payen et al. observed a shift in the Raman band from 380 to 385 cm^{-1} with an increase in Mo loading (Payen et al. [1987]); the authors attributed their shift (and broadening) to the lateral growth of MoS_2 particles without making reference to whether this 'crystallite size' effect is an actual broadening of the basal plane or merely side-to-side stacking of MoS_2 crystals. In our samples, it can be seen that the crystallite sizes do increase with an increase in decomposition pressure though the values do not correlate well with the above conclusion - our results show a general increase in basal plane size but the results for 7-20 in particular and all samples prepared at 500 psig in general show similar crystalline orders than the samples prepared at 1,000 psig; thus at present, it is inconclusive that the peak shift is due (at least primarily) to lateral growth. Since FT-IR spectra are of the bulk material, it could be argued that there is an increase in the concentration of weaker (longer), more reactive Mo-S bonds produced with increasing decomposition pressure. While this finding is intuitively unappealing as one would reasonably expect a shortening of bond lengths with increase in pressure (Pietosa et al. [2008]), a possible explanation is that pressure induces structural changes which in turn induce creation of new catalytically active sites. Though recent investigations of MoS_2 nanoclusters using direct space DFT calculations have found that an increase in S atom coordination of Mo atom results in increasing Mo-S bond lengths (McBride and Head [2009]), no such investigations were carried out in this study. Work is ongoing with a view to quantifying the interlayer rotations observed using XRD and the creation of new catalytically active sites and the unique FT-IR shifts observed through the preparation of more ATM-derived MoS_2 using more discreet changes in decomposition pressures.

One theory for the proposed catalytic activity of MoS_2 is that the presence of a coordinatively unsaturated site (CUS), with nearby SH groups, is a requirement for HDS (Lipsch and Schuit[1969]). This CUS is formed by removal of sulfur from a molybdenum atom and occurs in the presence of H_2 (Alonso et al. [1998]). Logically, therefore, the weaker the Mo-S bond the greater the concentration of sulfur vacancies that can be produced resulting in a higher activity (Nørskov et al. [1992]). One can expect an improved HDS activity of MoS_2 with an FT-IR peak at 378 cm^{-1} over MoS_2 with a stronger Mo-S bond at 385 cm^{-1}, when comparing similar activity amongst catalysts of comparable texture and with the same strength of Mo-S peak. This was observed in the present study as well, with the exception of specimen 7-1,000, which exhibited different textural properties from other specimens. From the conditions under consideration in this study, all catalysts prepared at high pressure exhibited the weaker Mo-S bond and the largest S_{BET} and the largest TPV were found in 5-1,000; thus, it was surmised to be the best catalyst prepared for HDS of DBT. While 5-1,000 did in fact have the highest activity in real terms (Table 2), as well as a function of TPV, it did not as a function of S_{BET} (Figure 7A,B, respectively). However, we have already shown that agglomeration occurs during

S_{BET} measurements, so a more accurate representation would be to compare activity against the theoretically calculated surface area (from XRD). On this basis, 5-1,000 has the highest activity per unit area of MoS_2.

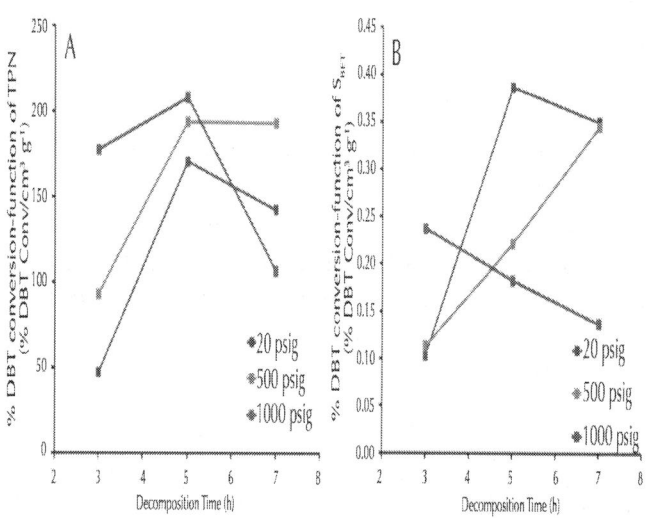

Figure 7: ATM-derived MoS2 activity as a function of TPV and S_{BET}. ATM-derived MoS2 activity as a function of total pore volume (TPV; shown in (A)) and BET-derived specific surface (S_{BET}; shown in (B)).

The increase in HYD:DDS ratio, as confirmed by chromatographic analysis, can be partially explained by a possible increase in the number of rim sites available due to interlayer rotation of the layers of MoS_2 in the stacks. Another possibility is that the reduced agglomeration from samples prepared at higher decomposition pressure simply exposes more rim sites to the DBT molecule. Of the samples which produce the hydrogenated DBT, 5-1,000 has one of the lowest HYD:DDS ratios (0.15) compared to say 3-500 (0.32). Thus, not only does 5-1,000 have a more active surface for HDS of DBT, but it also has a higher preference towards the DDS pathway compared to HYD which leads to the assumption that the exposure of rim sites for MoS_2 prepared from reductively decomposed ATM can be optimized by varying decomposition time and pressure.

CONCLUSIONS

We have shown herein the role played by both time and pressure for reductive thermal decomposition of ATM for the formation of MoS_2 and their subsequent impact on HDS activity. The influence of decomposition time serves to both enable an increase in surface area as well as an increased likelihood of sintering. Pressure increases slow down thermal decomposition by limiting gaseous by-product diffusion while at the same time facilitating the collapse of produced pores. Based on FT-IR vibrational energies, FT-IR spectral analysis shows that catalysts prepared at higher pressures have weaker Mo-S bonds (378 cm^{-1}) than those prepared at lower pressure (385 cm^{-1}). Evidence from XRD indicates both time and pressure induce interlayer rotation of stacks along both the a- and c-axes and that ATM decomposition pressure plays a role in determination of MoS_2 layer stacking.

Using XRD-derived values for crystallite sizes, theoretical surface area values were determined; the changes evident lead us to conclude that agglomeration is occurring. Optimal conditions were found producing a catalyst with weak Mo-S bonds, the largest (theoretical) surface area and largest pore volume which unsurprisingly resulted in the highest HDS activity of model dibenzothiophene. It was found that catalytic activity increased with a decrease in theoretical surface area, with 5-1,000 having the highest activity per unit area. Samples prepared at low pressures yielded no detectible HYD product, but this increased in samples prepared at higher decomposition pressure. An optimal system was found which gave the lowest HYD:DDS ratio, again 5-1,000. There is no doubt that there are other parameters which can impact both morphology and catalytic activity, such as ramp rates and variously substituted thiomolybdates; we have shown that the role of decomposition time and pressure for ATM-derived MoS_2 can in fact be a useful tool in optimizing catalyst synthesis with certain morphologies and activities.

AUTHOR'S CONTRIBUTIONS

JW: catalyst synthesis and characterization and manuscript writing; IB: catalyst testing; GEL: FTIR characterization; NDS: catalyst testing; SS: experimental design and execution; AT: XRD characterization; SAH: original idea and data interpretation; RVV: original idea and design of work; MSK: data interpretation and manuscript writing; SMA: overall supervision. All authors read and approved the final manuscript.

ACKNOWLEDGMENTS

The authors would like to acknowledge financial support from the Abu Dhabi Oil Refining Company (TAKREER) and from the Department of Chemical Engineering at The Petroleum Institute, Abu Dhabi, United Arab Emirates.

REFERENCES

1. Afanasiev P: Synthetic approaches to the molybdenum sulfide materials. Comptes Rendus Chimie 2008, 11(1–2):159-182. http://dx.doi.org/10.1016/j.crci.2007.04.009
2. Afanasiev P: The influence of reducing and sulfiding conditions on the properties of unsupported MoS2-based catalysts. J Catal 2010, 269(2):269-280. http://dx.doi.org/10.1016/j.jcat.2009.11.004
3. Alekseevskii NE, Dobrovol'skii NM, Eckert D, Tsebro VI: Investigation of critical currents of ternary molybdenum sulfides. J Low Temp Phys 1977, 29(5–6):565-572. doi:10.1007/bf00661547
4. Alonso G, Del Valle M, Cruz J, Petranovskii V, Licea-Claverie A, Fuentes S: Preparation of MoS2 catalysts by in situ decomposition of tetraalkylammonium thiomolybdates.

Catalysis Today 1998, 43(1–2):117-122. http://dx.doi.org/10.1016/S0920-5861(98)00140-0

5. Álvarez L, Berhault G, Alonso-Nuñez G: Unsupported NiMo sulfide catalysts obtained from Nickel/Ammonium and Nickel/Tetraalkylammonium thiomolybdates: synthesis and application in the hydrodesulfurization of dibenzothiophene. Catal Lett 2008, 125(1–2):35-45. doi:10.1007/s10562-008-9541-2

6. Berhault G, Cota Araiza L, Duarte Moller A, Mehta A, Chianelli R: Modifications of unpromoted and cobalt-promoted MoS2 during thermal treatment by dimethylsulfide. Catal Lett 2002, 78(1–4):81-90. doi:10.1023/a:1014910105975

7. Breysse M, Geantet C, Afanasiev P, Blanchard J, Vrinat M: Recent studies on the preparation, activation and design of active phases and supports of hydrotreating catalysts. Catalysis Today 2008, 130(1):3-13. http://dx.doi.org/10.1016/j.cattod.2007.08.018

8. Brunet S, Mey D, Pérot G, Bouchy C, Diehl F: On the hydrodesulfurization of FCC gasoline: a review. Appl Catal Gen 2005, 278(2):143-172. http://dx.doi.org/10.1016/j.apcata.2004.10.012

9. Camacho-Bragado GA, Elechiguerra JL, Olivas A, Fuentes S, Galvan D, Yacaman MJ:Structure and catalytic properties of nanostructured molybdenum sulfides. J Catal 2005, 234(1):182-190. http://dx.doi.org/10.1016/j.jcat.2005.06.009

10. Chianelli RR, Siadati MH, De la Rosa MP, Berhault G, Wilcoxon JP, Bearden R, Abrams BL:Catalytic properties of single layers of transition metal sulfide catalytic materials. Catalysis Rev 2006, 48(1):1-41. doi:10.1080/01614940500439776

11. Chianelli RR, Berhault G, Torres B: Unsupported transition metal sulfide catalysts: 100 years of science and application. Catalysis Today 2009, 147(3–4):275-286. http://dx.doi.org/10.1016/j.cattod.2008.09.041

12. Criado J, González M, Málek J, Ortega A: The effect of the CO_2 pressure on the thermal decomposition kinetics of

calcium carbonate. Thermochimica Acta 1995, 254(0):121-127. http://dx.doi.org/10.1016/0040-6031(94)01998-V

13. Daage M, Chianelli RR: Structure-function relations in molybdenum sulfide catalysts: the "Rim-Edge" model. J Catal 1994, 149(2):414-427. http://dx.doi.org/10.1006/jcat.1994.1308

14. De la Rosa MP, Texier S, Berhault G, Camacho A, Yácaman MJ, Mehta A, Fuentes S, Montoya JA, Murrieta F, Chianelli RR: Structural studies of catalytically stabilized model and industrial-supported hydrodesulfurization catalysts. J Catal 2004, 225(2):288-299. http://dx.doi.org/10.1016/j.jcat.2004.03.039

15. Devers E, Afanasiev P, Jouguet B, Vrinat M: Hydrothermal syntheses and catalytic properties of dispersed molybdenum sulfides. Catal Lett 2002, 82(1–2):13-17. doi:10.1023/a:1020512320773

16. Egorova M, Prins R: The role of Ni and Co promoters in the simultaneous HDS of dibenzothiophene and HDN of amines over Mo/ -Al2O3 catalysts. J Catal 2006, 241(1):162-172. http://dx.doi.org/10.1016/j.jcat.2006.04.011

17. Fedin VP, Kolesov BA, Mironov YV, Fedorov VY: Synthesis and vibrational (IR and Raman) spectroscopic study of triangular thio-complexes [Mo3S13]2− containing 92Mo, 100Mo and 34S isotopes. Polyhedron 1989, 8(20):2419-2423. http://dx.doi.org/10.1016/S0277-5387(89)80005-1

18. Iwata Y, Araki Y, Honna K, Miki Y, Sato K, Shimada H: Hydrogenation active sites of unsupported molybdenum sulfide catalysts for hydroprocessing heavy oils. Catalysis Today 2001, 65(2–4):335-341. http://dx.doi.org/10.1016/S0920-5861(00)00554-X

19. Klimov OV, Pashigreva AV, Fedotov MA, Kochubey DI, Chesalov YA, Bukhtiyarova GA, Noskov AS: Co–Mo catalysts for ultra-deep HDS of diesel fuels prepared via synthesis of bimetallic surface compounds. J Mol Catalysis

A Chem 2010, 322(1-2):80-89. http://dx.doi.org/10.1016/j.molcata.2010.02.020
20. Liang KS, Chianelli RR, Chien FZ, Moss SC: Structure of poorly crystalline MoS2 — a modeling study. J Non Cryst Solids 1986, 79(3):251-273. http://dx.doi.org/10.1016/0022-3093(86)90226-7
21. Lipsch JMJG, Schuit GCA: The CoO MoO3 Al2O3 catalyst: III. Catalytic properties. J Catal 1969, 15(2):179-189. http://dx.doi.org/10.1016/0021-9517(69)90022-0
22. McBride KL, Head JD: DFT investigation of MoS_2 nanoclusters used as desulfurization catalysts. Int J Quantum Chem 2009, 109(15):3570-3582. doi:10.1002/qua.22328
23. Mdleleni MM, Hyeon T, Suslick KS: Sonochemical synthesis of nanostructured molybdenum sulfide. J Am Chem Soc 1998, 120(24):6189-6190. doi:10.1021/ja9800333
24. Müller A, Jaegermann W, Enemark JH: Disulfur complexes. Coord Chem Rev 1982, 46:245-280 Nørskov JK, Clausen BS, Topsøe H: Understanding the trends in the hydrodesulfurization activity of the transition metal sulfides. Catal Lett 1992, 13(1-2):1-8. doi:10.1007/bf00770941
25. Oyumi Y, Brill TB: Thermal decomposition of energetic materials 22. The contrasting effects of pressure on the high-rate thermolysis of 34 energetic compounds. Combustion Flame 1987, 68(2):209-216. http://dx.doi.org/10.1016/0010-2180(87)90058-7
26. Payen E, Grimblot J, Kasztelan S: Study of oxidic and reduced alumina-supported molybdate and heptamolybdate species by in situ laser Raman spectroscopy. J Phys Chem 1987, 91(27):6642-6648. doi:10.1021/j100311a018
27. Peng Y, Meng Z, Zhong C, Lu J, Yu W, Yang Z, Qian Y: Hydrothermal synthesis of MoS2 and its pressure-related crystallization. J Solid State Chem 2001, 159(1):170-173. http://dx.doi.org/10.1006/jssc.2001.9146
28. Piermarini GJ, Block S, Miller PJ: Effects of pressure and temperature on the thermal decomposition rate and

reaction mechanism of.beta.-octahydro-1,3,5,7-tetranitro-1,3,5,7-tetrazocine. J Phys Chem 1987, 91(14):3872-3878. doi:10.1021/j100298a028

29. Pietosa J, Dabrowski B, Wisniewski A, Puzniak R, Kiyanagi R, Maxwell T, Jorgensen JD:Pressure effects on magnetic and structural properties of pure and substituted $SrRuO_3$. Phys Rev B 2008, 77(10):104410Polyakov M, van den Berg MWE, Hanft T, Poisot M, Bensch W, Muhler M, Grünert W:Hydrocarbon reactions on MoS2 revisited, I: activation of MoS2 and interaction with hydrogen studied by transient kinetic experiments. J Catal 2008, 256(1):126-136. http://dx.doi.org/10.1016/j.jcat.2008.03.007

30. Sánchez V, Benavente E, Lavayen V, O'Dwyer C, Sotomayor Torres CM, González G, Santa Ana MA: Pressure induced anisotropy of electrical conductivity in polycrystalline molybdenum disulfide. Appl Surf Sci 2006, 252(22):7941-7947. http://dx.doi.org/10.1016/j.apsusc.2005.10.011

31. Topsøe H, Clausen B, Massoth F: Hydrotreating catalysis. [10.1007/978-3-642-61040-0_1] In Catalysis, vol 11 Edited by Anderson J, Boudart M. Springer, Berlin Heidelberg; 1996, 1-269. doi:10.1007/978-3-642-61040-0_1

32. Walton RI, Dent AJ, Hibble SJ: In situ investigation of the thermal decomposition of ammonium tetrathiomolybdate using combined time-resolved X-ray absorption spectroscopy and X-ray diffraction. Chem Mater 1998, 10(11):3737-3745. doi:10.1021/cm980716h

33. Weber T, Muijsers JC, Niemantsverdriet JW: Structure of amorphous MoS3. J Phys Chem 1995, 99(22):9194-9200. doi:10.1021/j100022a037

34. Zhang F, Vasudevan PT: TPD and HYD studies of unpromoted and co-promoted molybdenum sulfide catalyst ex ammonium tetrathiomolybdate. J Catal 1995, 157(2):536-544. http://dx.doi.org/10.1006/jcat.1995.1317

Chapter 2

Effects of Multiwalled Carbon Nanotubes and Triclocarban on Several Eukaryotic Cell Lines: Elucidating Cytotoxicity, Endocrine Disruption, and Reactive Oxygen Species Generation

Anne Simon[1], Sibylle X Maletz[1], Henner Hollert[1, 2, 3, 4], Andreas Schäffer[1, 2, 3, 4], and Hanna M Maes[1]

[1]Institute for Environmental Research (Biology V), RWTH Aachen University, Worringerweg 1, Aachen 52074, Germany

[2]School of Environment, Nanjing University, Nanjing 210023, China

[3] Key Laboratory of Yangtze River Environment of Education Ministry of China, College of Environmental Science and Engineering, Tongji University, Shanghai 200092, China

[4] College of Resources and Environmental Science, Chongqing University, Chongqing 400715, China

ABSTRACT

To date, only a few reports about studies on toxic effects of carbon nanotubes (CNT) are available, and their results are often controversial. Three different cell lines (rainbow trout liver cells (RTL-W1), human adrenocortical carcinoma cells (T47Dluc), and human adrenocarcinoma cells (H295R)) were exposed to multiwalled carbon nanotubes, the antimicrobial agent triclocarban (TCC) as well as the mixture of both substances in a concentration range of 3.13 to 50 mg CNT/L, 31.25 to 500 µg TCC/L, and 3.13 to 50 mg CNT/L + 1% TCC (percentage relative to carbon nanotubes concentration), respectively. Triclocarban is a high-production volume chemical that is widely used as an antimicrobial compound and is known for its toxicity, hydrophobicity, endocrine disruption, bioaccumulation potential, and environmental persistence. Carbon nanotubes are known to interact with hydrophobic organic compounds. Therefore, triclocarban was selected as a model substance to examine mixture toxicity in this study. The influence of multiwalled carbon nanotubes and triclocarban on various toxicological endpoints was specified: neither cytotoxicity nor endocrine disruption could be observed after exposure of the three cell lines to carbon nanotubes, but the nanomaterial caused intracellular generation of reactive oxygen species in all cell types. For TCC on the other hand, cell vitality of 80% could be observed at a concentration of 2.1 mg/L for treated RTL-W1 cells. A decrease of luciferase activity in the ER Calux assay at a triclocarban concentration of 125 µg/L and higher was observed. This effect was less pronounced when multiwalled carbon nanotubes were present in the medium. Taken together, these results demonstrate that multiwalled carbon nanotubes induce the production of reactive oxygen species in RTL-W1, T47Dluc, and H295R cells, reveal no cytotoxicity, and reduce the bioavailability and toxicity of the biocide triclocarban.

BACKGROUND

The annual worldwide production of carbon nanotubes (CNT) surpassed the multimetric ton level and is expected to further increase [1]. Their structure gives them exceptional properties, which makes this material suitable for the use in composite materials, sensors, drug delivery, hydrogen storage fuel cells, and various environmental applications [2-4]. The probability of occupational and public exposure to CNT has significantly increased [5]. With this nanophase invasion of new materials and products into many aspects of life comes the need for increasing safety measures for exposure risks [6]. In October 2011, the European Union defined nanomaterials as natural, incidental, or manufactured materials containing particles, in an unbound state or as an aggregate or agglomerate, where 50% or more of the particles exhibited one or more external dimension in the size range of 1 to 100 nm [7]. Carbon nanotubes represent one of the most promising nanomaterials for various applications [8]. However, public concerns on the widespread use of these materials increase due to their close similarity to other toxic fibers regarding their high aspect ratio, reactivity, and biopersistence. Multiwalled carbon nanotubes (MWCNT) used in this study were the most highly produced CNT materials until 2012 [8]. A pilot plant with an annual capacity of 60 tons is since 2007 in an operation in southern Germany. Thus, knowledge on the toxic potential of MWCNT is required also regarding the very different nature of various types differing in flexibility or stiffness, varying in length and aspect ratio as well as having different contents of metal catalysts and surface properties. All MWCNT have a tubular structure with a high aspect ratio and between 2 and 30 concentric cylinders with outer diameters commonly between 30 and 50 nm. The small size and the high surface area define the chemical reactivity of CNT and induce changes in permeability or conductivity of biological membranes [9]. Therefore, engineered CNT may pose health risks because of their ability to reach every part of the organs and tissues and their interaction with cellular functions. The primary risk of these materials may come from their ability to enter cells, which may cause damage to plants, animals, and humans [10-13]. Important characteristics are the surface chemistry and purity of CNT. For MWCNT synthesized using a metal catalyst, the toxicity may be the combined effect of the MWCNT themselves and an oxidative stress response to the residual metal catalyst [14] typically amounting to less than about 5 wt.%. This complicates clear determination of pure MWCNT toxicity. Despite these concerns, very few studies have been simultaneously conducted with various human cell lines to assess the health effects of different CNT. At present, there is no global agreement about the risk of CNT on human health [15].

Previous researchers have explored the toxicity of carbon nanomaterials to animal and human cells [16-20]. It was suggested that the toxicity of carbon nanomaterials may also be caused by sorption of toxic substances to their surface [21-23]. Therefore, knowledge of toxic compound adsorption by carbon nanomaterials is critical and useful for risk assessment of these nanomaterials because in the environment, both nanomaterials and chemical pollutants, are present as complex mixtures.

CNT are carbonaceous adsorbents with hydrophobic surfaces that exhibit strong adsorption affinities to organic compounds [24-30]. Thereby, a combination of chemical and physical interactions play a major role for adsorption processes. CNT have uniform structural units but are prone to aggregate, forming bundles of randomly tangled agglomerates because of the strong van der Waals forces along the length axis [31]. The outermost surface, inner cavities, interstitial channels, and peripheral grooves of CNT constitute four possible sorption sites for organic compounds [30]. Nanotechnology has initiated different types of nanomaterials to be used in water technology in recent years that can have promising outcomes. Nanosorbents such as CNT have exceptional adsorption properties and can be applied for removal of heavy metals, organics, and biological impurities [28,32]. CNT, as adsorbent media, are able to remove heavy metals such as Cr^{3+}[33], Pb^{2+}[34], and Zn^{2+}[35], metalloids such as arsenic compounds [36], organics such as polycyclic aromatic organic compounds (PAH) [24,29], pesticides [37], and a range of biological contaminants including bacteria [38-40], viruses [41,42], cyanobacterial toxins [43,44] as well as natural organic matter (NOM) [45-47]. The success of CNT as an adsorbent media in the removal of biological contaminants, especially pathogens is mainly attributed to their unique physical, cytotoxic, and surface functionalizing properties [28].

To date, many studies on the safety of different CNT materials have been conducted but the results are often controversial and depending of the species of the applied CNT. A wide range of results from in vitro studies, dealing with MWCNT, has been reported. On the one hand, MWCNT decreased cell viability and induced apoptosis [48,49], whereas minimal to no decrease of cell viability was observed [50]. One explanation of this controversy is the type of cells used. Additional explanations are that MWCNT are produced by different processes, tested with varying dispersion methods, and that their life cycle may confer changes in their surface characteristics and reactivity. For example, in some studies, the presence of metal trace impurities explains demonstrated toxicity and reactive oxygen species (ROS) production [50], whereas in other cases, no such effects were reported [51]. Nevertheless, it is recognized that nanoparticles produce ROS[50,52] inside and outside the cell, which has to be considered as one of the key factors for toxicological effects [6]. Hence, further evaluation and characterization of their toxic potential and other effects on cells like cytotoxicity, endocrine disruption, and the production of ROS, which can result in cell damage, is of highest concern.

Relatively little research has been conducted examining biocidal components of personal care products, as for example triclocarban (TCC), although such products are continually released into the aquatic environment and are biologically active and some of them persistent [53]. Therefore, they are detected often and in rather high concentrations in the environment [53]. TCC is a high-production volume chemical [54] that is widely used as an antimicrobial compound [53,55]. It is able to adsorb on the cell membrane and to destroy its semi-permeable character, leading to cell death [56]. In the U.S., the annual production of TCC in 2002 added up to 500 metric tons [57,58]. The primary route for TCC to enter the environment is through discharge of effluent from wastewater treatment plants and disposal of solid residuals on land [55,58]. Due to its lipophilicity (log Kow 4.9 [59]), TCC has an affinity to adsorb to organic matter [60]; therefore, over 70% of the initial mass has been found to be adsorbed to sludge [61,62]. TCC has been detected at microgram per liter levels in waterways in the United States and Switzerland, indicating extensive contamination of aquatic ecosystems [54,63,64]. TCC was chosen in this study for its widespread use, toxicity [58], bioaccumulation potential [65,66], environmental persistence, and endocrine effects [67].

As TCC is used since 1957 in huge amounts [53], and MWCNT is supposed to reach the amount of a large scale production, both substances might involuntarily occur together in the environment.

This study aimed to provide new information on toxicity of TCC and nanotoxicity of MWCNT as well as the mixture of both substances by using three different eukaryotic cell lines. Key questions were to get more information about the cytotoxicity of MWCNT and the estrogenic potential of TCC as well as the potential of MWCNT to generate ROS in cell lines. Especially, the interaction of MWCNT and TCC poses a major question in the present study, if one of them is more or less toxic when cells are exposed to mixtures of both.

As many studies already showed that CNT are toxic for different cell lines [5,9], we investigated cells by determination of cytotoxicity in the neutral red retention (NR) assay and the 3-(4,5-dimethylthiazol-2-yl)-2,5-diphenyl tetrazolium bromide (MTT) assay [68] to verify whether MWCNT showed a toxic potential for the used cells, namely RTL-W1, T47Dluc, and H295R.

A combination of cytotoxicity assays, particularly the NR and MTT assay, was preferred in many studies [69-71], as this would increase the reliability of the results obtained. Furthermore, mechanism-specific endpoints, such as estrogenic effects and alterations of the steroid synthesis were analyzed by using the estrogen receptor-mediated chemical-activated luciferase gene expression (ER-Calux) assay [72] and the H295R steroidogenesis assay (H295R)[73,74], respectively. The evaluation of the endocrine activity in wastewater samples could already been proven by using these assays [75-78]. As previously reviewed by Hecker and Hollert [79], results of several studies indicated that a combined use of receptor-mediated and non-receptor-mediated methods is necessary to enable objective assessment of endocrine potential in complex samples. Additionally, Grund et al. [80] demonstrated that the combination of receptor-mediated and non-receptor-mediated assays such as the ER Calux and the H295R was appropriate for a holistic evaluation of potential endocrine activity of complex environmental samples.

The measurement of cellular reactive oxygen species was investigated by using the fluorescent dye 2',7'-dichlorodihydrofluorescein diacetate (H_2DCF-DA) assay [81].

METHODS

Chemicals

The test substance 3,4,4'-trichlorocarbanilide was purchased from Sigma Aldrich (St. Louis, MO, USA) and had a purity of 99% (CAS:101-20-2). Multiwalled carbon nanotubes (Baytubes C150P, >95% purity) were provided from Bayer MaterialScience (Bayer AG, Leverkusen, Germany). The used concentrations of both materials in the different test systems were based on limit tests and not higher than the dispersibility of CNT or solubility of TCC.

Cell Cultures

RTL-W1 Cells

The rainbow trout liver cell line (RTL-W1) [82] was grown in L15-Leibovitz medium (Sigma-Aldrich) supplemented with 9% fetal bovine serum (FBS, Biowest, Logan, UT, USA) and penicillin/streptomycin (10,000 E/mL; 10,000 µg/mL in 0.9% NaCl, Sigma-Aldrich) in 75-cm^2 flasks (Techno Plastic Products (TPP), Trasadingen, Switzerland) at 20°C in darkness according to the protocol detailed in Klee et al. [83].

T47Dluc Cells

The human T47Dluc breast adenocarcinoma cells were obtained from BioDetection Systems BV (Amsterdam, the Netherlands) and were cultured in Dulbecco's modified Eagle medium/nutrient mixture F-12 (DMEM/F12) with phenol red (Gibco, Grand Island, NY, USA) supplemented with sodium bicarbonate (Sigma-Aldrich), MEM 100× (Gibco), penicillin/streptomycin solutions (Gibco) and 7.5% fetal bovine serum (FBS) according to the methods details in Maletz et al. [84]. T47Dluc cells were cultured at 37°C, 7.5% CO_2, and maximum humidity.

H295R Cells

The human adrenocarcinoma cells (H295R) were obtained from the American Type Culture Collection (ATCC; Manassas, VA, USA) and were grown in 75-cm^2 flasks with 8 mL supplemented medium at 37°C with a 5% CO_2 atmosphere as described previously [73,85].

Nanoparticles Suspension

Test suspensions of 1 to 100 mg/L of MWCNT were prepared by ultrasonication of the raw material with a microtip (70 W, 0.2" pulse and 0.8" pause; Bandelin, Berlin, Germany) in distilled water for 10 min. Transmission electron microscopy (TEM) images showed the presence of small agglomerates and individual nanotubes in the medium (Figure 1).

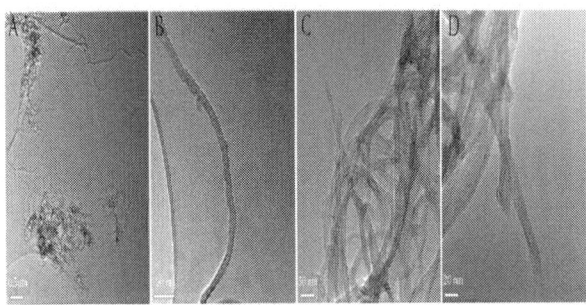

Figure 1: TEM pictures of MWCNT. Agglomerates (A), single nanotubes (B), and tubes sticking out of the agglomerates (C, D) visualized by transmission electron micrographs of sonicated MWCNT in distilled water.

Cytotoxicity Assays

For determining the effect of particles on cell viability, different assays were used. Potential interferences of MWCNT and the fluorescence measurement were prevented by using black microtiter plates.

Neutral Red Retention Assay

The neutral red retention (NR) assay was performed according to Borenfreund and Puerner [86] with slight modifications as detailed in Heger et al. [87] by using RTL-W1 cells. Briefly, 4×10^5 cells were seeded into each well (except for the blanks) of a 96-well microtiter plate (Nunc) and directly treated in triplicates with the particle suspensions. To guarantee optimal culture conditions, cells were exposed in a 1:1 mixture of MWCNT suspension or TCC solution and double-concentrated L15-Leibovitz medium, resulting in final MWCNT-concentrations of 3.13 to 50 mg CNT/L and TCC concentrations of 7.8 to 10×10^3 mg/L. After incubation for 48 h at 20°C in the dark, the sample solution was discarded, and each well was rinsed with 100 µL phosphate-buffered saline (PBS) to remove any excess medium. One hundred microliters of a 0.005% neutral red solution (2-methyl-3-amino-7-dimethylaminophenanzine, Sigma-Aldrich) was added to each well except for the blanks. After an incubation time of 3 h at 20°C in darkness, the amount of extracted NR was determined by absorption measurement at 540 nm and a reference wavelength of 690 nm using a microtiter plate reader (Infinite M200, Tecan Instruments, Männedorf, Switzerland). Thereafter, concentrations resulting in cell vitality of 80% were calculated and identified as NR_{80} values according to Heger et al. 2012 [87]. For detection of significant differences, the t test following square root transformation was performed using SigmaPlot 12. Results are given as relative values to the untreated control in percent.

MTT Assay

The cell viability was evaluated by the reduction of water soluble 3-(4,5-dimethylthiazol-2-yl)-2,5-diphenyl tetrazolium bromide (MTT, Sigma Aldrich) to water-insoluble formazan crystals by mitochondrial dehydrogenase [88]. The amount of the formed blue formazan is proportional to the amount of viable cells [89], and the absorbance was measured at 492 nm using a microtiter plate reader (Tecan).

H295R Cells

The exposure of H295R cells was conducted according to the methods of Hecker et al. [73,74]. In brief, 1 mL of cell suspension, at a concentration of 2.5×10^5 H295R cells/mL, was added to each well of a 24-well microtiter plate and cells were allowed to attach for 24 h. Cells were treated in triplicate with a 1:1 mixture of the MWCNT suspension and/or TCC solution and double-concentrated medium, resulting in final concentrations of 3.13 to 50 mg CNT/L and 31.25 to 500 µg TCC/L for 48 h as well as the two reference substances forskolin and prochloraz (quality control plate).

The plates were checked for cytotoxicity and contamination after 24 h of exposure. The culture supernatants were removed and frozen at -80°C for later analysis of alterations in steroid synthesis in the enzyme-linked immunosorbent assay (ELISA) assay. Cells were rinsed with 600 µL PBS per well. Then, 400 µL of a freshly prepared MTT (thiazolyl blue tetrazolium bromide, ≥ 97.5% TLC) solution at 500 µg/mL was added to each well and incubated for 30 min at 37°C and 5% CO_2 in air atmosphere. The MTT solution was discarded, and 800 µL DMSO was added to each well in order to lyse the cells. Plates were finally placed on a horizontal shaker for 10 to 15 min before measuring the absorbance. Results are given as relative values to the solvent control in percent.

T47Dluc Cells

The MTT assay was performed according to Mosmann [90]. In brief, T47Dluc cells were seeded into a 96-well microtiter plate (TPP) at a density of 1×10^4 cells per well. After 24 h of pre-incubation, the old medium was removed and cells were treated with a 1:1 mixture of the MWCNT suspension and/or TCC solution and double-concentrated medium. A serial dilution resulted in five concentrations of the MWCNT suspension and TCC solution and a solvent control were applied to each plate. For each concentration, three wells were foreseen. The exposure medium was removed, and the absorbance was measured after adding the freshly prepared MTT solution (500 µg/mL, Sigma-Aldrich) with a luminescence counter (Tecan) at 492 nm.

For both cell lines (H295R and T47Dluc), concentration-response curves were fitted with a non-linear 'log(agonist) vs. response - variable slope' regression using GraphPad Prism 5 as detailed in Heger et al. [87].

ER Calux

The ER Calux assay with stably transfected T47Dluc human breast cancer cells was developed by Legler et al. [72] and was conducted in this study according to the detailed protocol given in Maletz et al. [84]. T47Dluc cells/mL (10×10^4), resulting in a density of 1×10^4 cells per well, were plated into 96-well microtiter plates in medium (DMEM/F12 free of phenol red supplemented with sodium bicarbonate, MEM 100×, and fetal calf serum) and incubated for 24 h at 37°C (7.5% CO_2, 100% humidity). After this time, the assay medium was renewed, and the cells were incubated for another 24 h. Then, a 1:1 mixture of the MWCNT suspension and/or TCC solution and double-concentrated medium replaced the medium by using a serial dilution resulting in five concentrations. All concentrations of the test compound and the positive control (E2) as well as blanks (DMSO) and solvent control (EtOH) were introduced to each plate in triplicate. After 24 h of exposure, the plates were checked for cytotoxicity and contamination and the medium was removed. Following the addition of a mixture of 1:1 of PBS and steady light solution (PerkinElmer Inc., Waltham, MA, USA), the plates were incubated on an orbital shaker in darkness for 15 min. Luminescence was measured using a plate reader (Tecan). The luciferase activity per well was measured as relative light units (RLU). The mean RLU of blank wells was subtracted from all values to correct for the background signal. The relative response of all wells was calculated as the percentage of the maximal luciferase induction determined for E2 [91]. Only suspensions that did not cause cytotoxicity were used for quantification of the response.

Enzyme-linked Immunosorbent Assay

For quantification of hormone production by H295R cells, the protocol given by Hecker et al.[73,74] was used. To ensure that modulations in hormone synthesis were not a result of cytotoxic effects, viability of the cells was assessed with the MTT bioassay [90] before initiation of exposure experiments. Only non-cytotoxic concentrations (>80% viable cells per well) were evaluated regarding their potential to affect steroid genesis [80]. In brief, H295R cells were exposed as described above. The frozen medium was thawed and extracted using liquid extraction with diethylether as described previously in Maletz et al. [84]. The amount of 17 -estradiol (E2) was determined in an enzyme-linked immunosorbent assay (ELISA) assay (Cayman Chemicals, Ann Arbor, MI, USA) [80].

Measurement of Cellular ROS

The production of reactive oxygen species in RTL-W1, T47Dluc, and H295R cells were measured using the fluorescent dye 2′,7′-dichlorodihydrofluorescein diacetate (H_2DCF-DA) as previously described [50,81,92-95]. This dye is a stable cell-permeant indicator which becomes fluorescent when cleaved by intracellular esterases and oxidized by intracellular hydroxyl radical, peroxynitrite, and nitric oxide [92]. The intensity of fluorescence is therefore proportional to the amount of reactive oxygen species produced in cells. RTL-W1, T47Dluc, and H295R cells were charged as explained above, except for that H295R cells were seeded in 96-well plates as well. After an exposure time of 24 or 48 h, the medium was discarded, cells were washed three times with PBS because black particles strongly reduced the fluorescence signal, and 100 µL of H_2DCF-DA (final concentration of 5 µM in PBS) was added to each well. Subsequently, the plates were incubated for 45 min at room temperature on a horizontal shaker in darkness. Fluorescence at excitation and emission wavelengths of 485 and 530 nm, respectively, was measured with a microtiter plate reader (Tecan).

Statistical Methods

Statistical analyses were carried out with SigmaPlot 12. Results are presented as mean±standard deviation (SD). To enhance the comparability of the assays, the results were normalized to the average value of the solvent controls (SC) and are expressed as percent change or fold change relative to the SC. Prior to conducting statistical analyses, all data were checked for normality and homogeneity of variance using the Kolmogorov-Smirnov and Levene's test. A one-way analysis of variance (ANOVA) followed by Dunnett's post hoc test was used to determine treatments that differed significantly from the SC for data fulfilling the parametric assumptions. Otherwise, the non-parametric Kruskall-Wallis test followed by Dunn's post hoc test was used. For the detection of significant differences in cytotoxicity assays, the t test following square root transformation was performed. Differences were considered significant at $p < 0.05$.

RESULTS

Cytotoxicity

Neutral Red Retention Assay

An NR_{80} value (concentrations resulting in 80% viability of the RTL-W1 cells) of 2.1 mg/L was obtained for the biocide TCC (Figure 2). The exposure of cells to MWCNT at concentrations ranging between 0.78 and 50 mg/L and to the mixture of CNT and TCC (0.39 to 25 mg CNT/L +1% TCC; percentage relative to CNT concentration) did not result in cytotoxicity.

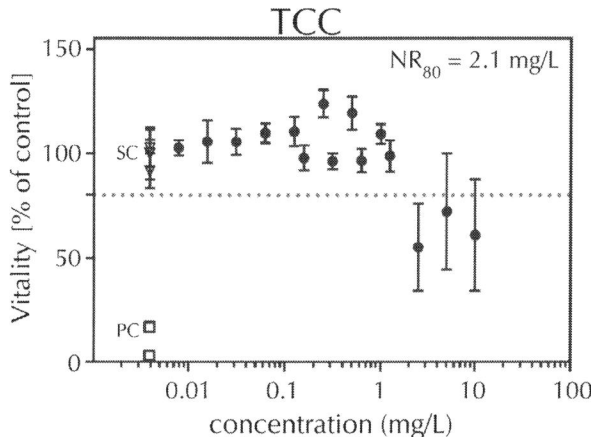

Figure 2: Cytotoxic effects of TCC in the NR assay. Cytotoxicity of TCC assessed in the neutral red retention assay with RTL-W1 cells. Dots represent the mean of three independent exposure experiments with three internal replicates and are given in percent of the viability of the control. The whiskers show the standard deviation of the mean; PC, positive control (3,5-dichlorophenol); SC, solvent control (EtOH); the dashed line marks the threshold of 80%.

Concentrations of TCC in the subsequently ROS assay were kept below 0.5 mg/L, i.e., below the NR_{80} value of 2.1 mg/L.

MTT Assay

In addition to the testing of RTL-W1 cells, cytotoxicity was assessed for T47Dluc cells and H295R cells in the MTT assay.

All concentrations of MWCNT (0.5 to 50 mg/L), TCC (31.25 to 500 µg/L), and the mixture of both substances (1.56 mg CNT/L + 15.6 µg TCC/L to 25 mg CNT/L + 250 µg TCC/L, i.e., CNT + 1% TCC) did not result in cytotoxicity in T47Dluc cells (data not shown). The results of the MTT cell viability assay with H295R cells are presented in Figure 3. The percentage of viable cells relative to the ethanol (EtOH) control is plotted against the respective sample concentration. The highest concentration of TCC (500 µg/L) revealed cytotoxicity after the exposure to H295R cells. In combination with

CNT, lower cytotoxicity of the biocide was observed although the same concentration of TCC was applied to the cells (Figure 3). The lower cytotoxicity of the mixture testing was not significantly different from the exposure to TCC alone. MWCNT-treated cells showed no cytotoxicity after exposure to concentrations between 3.13 and 50 mg CNT/L (data not shown).

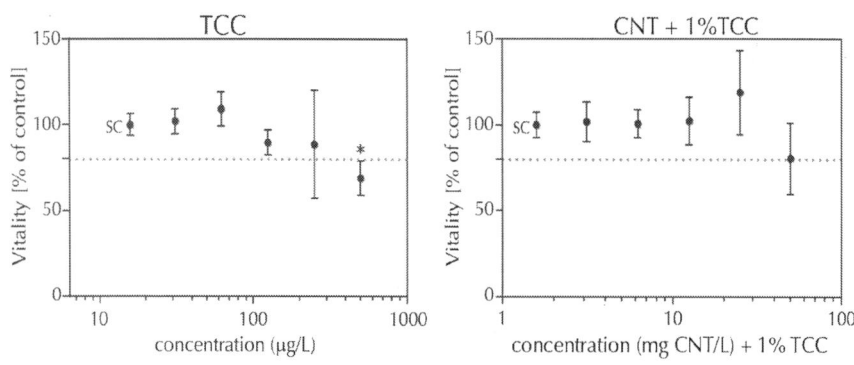

Figure 3: Cytotoxicity of TCC and its mixture with CNT in the MTT assay with H295R cells. Cytotoxicity of TCC and a mixture of CNT with 1% TCC (percentage relative to CNT concentration) as assessed in the MTT cell viability assay with H295R cells. Percent of viable cells after 48 h of exposure are given compared to the solvent control. Dots represent the mean of four independent exposure experiments with three internal replicates each. Error bars, standard deviation; SC, solvent control. The dashed line marks the threshold of 80%.

ER Calux Assay

Estrogenic activities were determined in CNT suspensions, TCC dilutions, and mixture of both substances using the ER Calux assay. Figure 4A shows that CNT had no estrogenic effect in the range of 3.13 to 50 mg CNT/L. Interestingly, a decrease of luciferase activity by high concentrations of the biocide TCC can be seen in Figure 4B. Cytotoxicity could be excluded for the concentrations used as shown in the MTT assay with T47Dluc cells. The antiestrogenic potential of TCC was reduced when cells were exposed to the mixture of CNT and 0.5% TCC (Figure 4C). This effect was not observed after application of CNT including 1% TCC (Figure 4D).

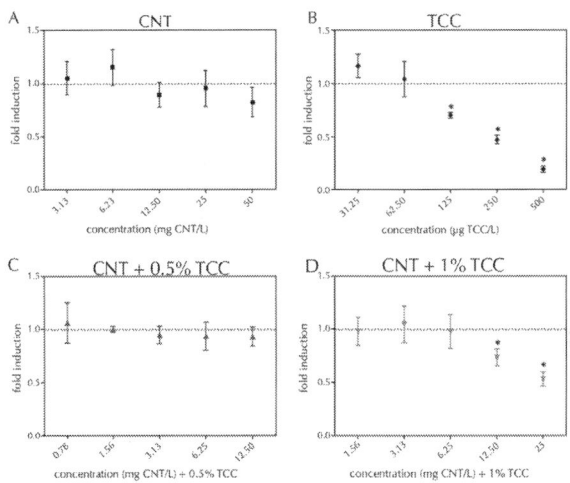

Figure 4: Estrogenic disruption in the ER Calux assay with T47Dluc cells. Estrogenic activity given as luciferase induction relative to solvent control (=1, dashed line) in the ER Calux assay plated in 96-well plates. T47Dluc cells were treated with CNT (A), TCC (B), and mixture of both (CNT+0.5% TCC (C), 1.56 mg CNT/L+7.80 µg TCC/L to 25 mg CNT/L+125 µg TCC/L; CNT+1% TCC (D), 1.56 mg CNT/L+15.60 µg TCC/L to 25 mg CNT/L+250 µg TCC/L). Dots represent means of two independent exposure experiments with three internal replicates each. Error bars, standard deviation; *statistically significant from the EtOH control in repeated measures ANOVA on Ranks with Dunn›s post hoc and $p<0.05$.

Alterations of Steroid Synthesis in H295R Cells

CNT did not have a pronounced effect on hormone production of 17β-estradiol (E2) in H295R cells. E2 levels were all in the range of the negative control. Also, after exposure to TCC concentrations, the hormones were at the level of the EtOH control. Mixture of

CNT and TCC did not significantly alter production of E2 in H295R cells in the range of 1.56 mg CNT/L + 15.6 µg TCC/L to 25 mg CNT/L + 250 µg TCC/L.

Measurement of Cellular ROS

Effects of MWCNT and TCC on radical formation were assessed by measuring intracellular ROS in RTL-W1, T47Dluc, and H295R cells. Compared to the EtOH control, no significant difference in the ROS generation by TCC and the combination of MWCNT and TCC in all three cell lines was observed. In MWCNT-treated cells, however, a much higher ROS production than that in the controls was measured. The ROS content was 1.8, 2.9, and 4.7 times higher compared to the control levels in RTL-W1 cells, 1.5, 1.9, and 3.2 times higher than in T47Dluc cells, and 1.2, 1.4, and 2.2 times higher than in H295R cells following incubation with CNT at 12.50, 25, and 50 mg/L, respectively (Figure 5).

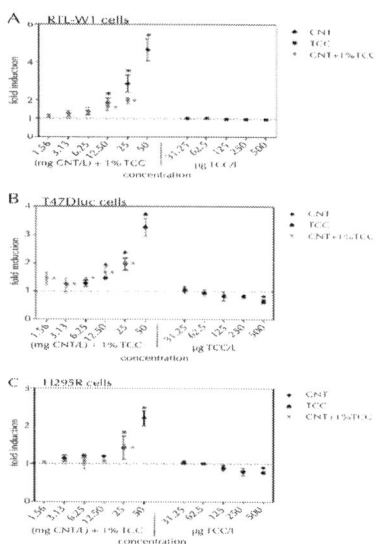

Figure 5: Generation of ROS in RTL-W1, T47Dluc, and H295R cells. ROS generated in RTL-W1 (A), T47Dluc (B), and H295R (C)

cells exposed to MWCNT, TCC, and mixture of both substances (1% TCC, with respect to the concentration of CNT). The intensity of H2DCF-DA was measured in cell lysates and normalized to negative/solvent control (=1, dashed line). Data are expressed as mean ± standard deviation of three independent exposure experiments with three internal replicates each. *Statistically significant from the negative control in repeated measures ANOVA on ranks with Dunn's post hoc and $p < 0.05$. The lowest observed effect concentration (LOEC) was 12.50 mg/L for RTL-W1 and T47Dluc cells, with a no observed effect concentration (NOEC) of 6.25 mg/L. For H295R cells, higher LOEC and NOEC were determined amounting to 25 and 12.5 mg CNT/L, respectively.

DISCUSSION

Multiwalled Carbon Nanotubes

In the case of long and stiff CNT, it has been argued that analogous mechanisms to those of other fibrous particles such as asbestos exist [96,97], which may penetrate the lung and persist in the tissue. The biopersistence, large aspect ratio, and fibrogenic character of CNT are important features that may cause adverse health effects. Other mechanisms include hydrophobic contact, through which nanoparticles may interrupt cell membranes, disturbing surface protein receptors[98]. Uptake of nanofibers by human macrophages sized smaller than the length of the nanotubes - a process defined as frustrated phagocytosis - has been shown by backscatter scanning electron microscopy [13]. Overall, nanomaterial size and composition plays a distinct role in the cellular response. In addition, this response is variable between cell types and is likely related to the physiological function of the cell types [95].

However, in our study, flexible multiwalled CNT were investigated for which less concern of their toxic potential has been expressed [99].

Cytotoxicity

Exposure of RTL-W1, T47Dluc, and H295R cells to 50 mg CNT/L for 24 or 48 h did not induce acute cell toxicity. This is the first study reporting data of cytotoxicity tests with Baytubes using these three cell lines. Several authors have shown that other types of CNT were cytotoxic to different lung epithelial cell lines [100-102], to human astrocyte D384 cells [100], to skin keratinocyte cells, lung cells, T4 lymphocytes [103], and human epidermal keratinocytes [18]. However, in a recent study, Thurnherr et al. [8] also showed that the same type of industrially produced MWCNT had no effect to another cell line. Contradiction to different effects observed in this study and in many other publications might be explained by differences in the CNT material used (metal contaminants, structural defects, size, stiffness, MWCNT vs. SWCNT) and by cell line dependency [8,92]. More likely, positive results are often only due to very high concentrations, which already elicit cytotoxic responses [104,105] or might interfere with the test systems used [106]. The hydrophobic nature of CNT is a general problem when working with these materials not only concerning the generation of stable suspensions that can be applied to the cultures but also for potential interference with the assay due to their high propensity to stick to various molecules or cells [107,108]. For this reason, we used no detergents to prevent MWCNT aggregation during the experiments. The exclusion of such interference with the test systems as well as thorough material characterization is therefore a prerequisite for each study to allow the comparison of results obtained from different researchers [109].

ROS Generation

Main effects of CNT seem to be due to oxidative stress, which triggers inflammation via the activation of oxidative stress-responsive transcription factors [110].

The highest intracellular ROS production could be observed in MWCNT-treated RTL-W1 cells, which was up to five times higher than control levels. A LOEC of 12.5 mg CNT/L was determined. They were followed by MWCNT-treated T47Dluc cells, in which up to three times more ROS was produced compared to the control. The lowest generation of ROS was observed in H295R cells with up to two times higher ROS levels compared to the control level with a LOEC of 25 mg/L.

ROS production can be partially inhibited by metal chelators, indicating that metal components (nickel, iron, yttrium) of CNT are able to contribute to the oxidant response observed [105]. CNT can contain relatively high concentrations of metals as impurities (e.g. 30%), which can contribute to their toxicity. In contrast, purified carbon nanotubes with no bioavailable metals were shown to decrease local oxidative stress development [111], suggesting that similar to fullerenes, ROS may be 'grafted' at the surface of CNT via radical addition due to their high electron affinity [110]. Barillet and coworkers came also to the conclusion that CNT induced the same level of ROS whatever their length and purity was [92]. They suggested that intracellular ROS production induced by CNT exposure refers to more complex mechanisms than simple redox reactions if we consider the fact that CNT are less accumulated than metal oxide nanoparticles [92].

Ye et al. [102] suggested that ROS and the activation of the redox-sensitive transcription factor NFkappaB were involved in upregulation of interleukin8 in A549 cells exposed to MWCNT. Yang et al. [112] found that CNT induced significant glutathione depletion, malondialdehyde increase, and ROS generation in a dose dependent manner. Pulskamp et al. [50] failed to observe any acute toxicity using the WST-1 assay in cultured rat NR8383 macrophages and A549 cells on viability and inflammation upon incubation with CNT. But they indicated a dose-dependent decrease of the mitochondrial enzyme activity (MTT assay) after 24 h of exposure, similar to the results seen before in other published studies [16, 17, 113] and detected a dose and time dependent increase of intracellular ROS [114]. ROS induction was also observed by exposure to carbon black [115]. Some doubt on the evaluation of MTT toxicity assays were expressed by Wörle-Knirsch et al. [116]

because they demonstrated that MTT formazan interacts with CNT interfering with the basic principle of the assay. The authors strongly suggest verifying cytotoxicity data with an independent test system as we did by using different test systems.

A key finding in our study was that ROS generation in three cell lines (RTL-W1, T47Dluc, and H295R) went up in 45 min even in a low dose of incubation group (3.13 mg/L), which was 1.2 times higher than in the controls. Chen et al. [114] assumed that ROS generation came out much earlier than other phenotypes including oxidative stress and cytotoxicity. This might be the reason why other studies in which ROS was measured after more than 4 h exposure to CNT showed inconsistent results [50,117-119]. Several studies [112,120] concluded that cytotoxicity can be attributed to oxidative stress. Interestingly, no cytotoxic effect was found in this study in three different MWCNT-treated cells, although generation of ROS was observed in all cell lines used.

Similar experiments to determine the ROS generation in RTL-W1 cells were performed using multilayer graphene flakes (synthesized by thermal reduction of graphitic oxide at the Federal Institute for Materials and Research and Testing BAM, Berlin) as non-nanomaterial (data not shown). Thereby, same increases of ROS generation were observed up to concentrations of 12.5 mg/L. Whereas, 1.5 times lower increases could be observed for both 25 and 50 mg/L compared to the MWCNT treatment. This lead us to the conclusion that the impurities of metal catalysts (cobalt) are not responsible for the increased production of ROS and such effects may be due to the nanostructure of these materials. Our findings are in accordance with other studies where intracellular ROS generation could be determined by using pristine graphene-treated murine RAW 264.7 macrophages [121], few-layer graphene (3 to 5 layers)-treated PC12 cells [122], and graphene oxide-treated human lung epithelial cells [123] in a time- and dose-dependent manner. However, Creighton et al. [124] showed that graphene-based materials have significant potential to interfere with in vitro toxicity testing methods, such as the H_2DCF-DA assay, through optical and adsorptive effects at toxicologically relevant doses (less than 10 to 100 mg/L). They could also show that the removal of the nanomaterial by washing can remove optical interferences. Depending on the graphene material, the washing step can lead to accurate data (e.g., for graphene oxide) or to underreporting of ROS as few-layer graphene (3 to 5 layers) adsorbs and quenches the H_2DCF-DA dye in a manner that depends on surface area [124]. Optical interferences can be excluded for the present study because the cell lines were washed accurately with PBS, but the adsorptive effect is still unclear and may lead to underestimate the production of ROS generation. Still, significant ROS production was observed in all three tested cell lines for the first time after exposure to Baytubes.

Triclocarban

Cytotoxicity

There is very limited information concerning the cytotoxic actions of TCC in mammalian cells, although these actions have been examined, to some extent, in aquatic and terrestrial organisms[125-127]. Morita et al. [126] showed no cell lethality after the incubation of rat thymocytes with TCC at concentrations ranging from 30 to 500 nM for 1 h. The incubation with TCC at concentrations ranging from 10 to 1 µM for 1 h did not affect the viability of rat thymocytes [128]. Another study by Kanbara et al. [129] showed an increase in cell lethality when rat thymocytes were incubated with 10 µM TCC. In the present study, a cytotoxic effect to treated RTL-W1 cells was already observed at concentrations above 4 µM TCC. Both human cell lines (T47Dluc, H295R) showed no cell lethality when exposed up to 1.6 µM TCC. These results are in agreement with the open literature [128,129].

Estrogenic Activity

As shown in Figure 4, a decrease of luciferase activity in the ER Calux assay was determined after exposure to high TCC concentrations (1.6 µM). Downregulation of estrogen receptors (ER) or other mechanisms of negative feedback may cause this decrease [130]. TCC did not significantly alter the production of E2 in H295R cells up to a concentration of 1.6 µM determined in the ELISA assay.

Ahn et al. [54] observed weak ER activity of TCC at concentrations of 1 and 10 µM. They also found that in the presence of estrogen or testosterone (T), TCC enhanced the actions of these hormones. A cell-based androgen receptor-mediated bioassay with TCC was investigated by Chen et al. [67]. Neither cytotoxicity nor the competition between TCC and testosterone for binding sites could be observed in their studies. However, TCC did amplify testosterone-induced transcriptional activity both in a time- and dose-dependent manner [67]. Altogether, the results suggest that the effects seen with TCC in luciferase-based transactivation assays are due to interference with firefly luciferase, rather than due to causing of the ER or the androgen receptor (AR) [131]. Similar false positives have been reported in previous high-throughput screens [132]. A recent screen of the NIH Molecular Libraries Small Molecule Repository identified 12% of the 360,864 molecules to be inhibitors of firefly luciferase [133]. In some cases, inhibition paradoxically resulted in an increase of the luminescence signal, probably because of enzyme stabilization [134]. Such a mode of action is also supported by the PubChem Bioassay Database (http://pubchem.ncbi.nlm.nih.govwebcite), which quotes a preliminary EC50 value of 8.9 µM TCC for the inhibition of luciferase.

The focus of the present study was to get more information about the biocide in cell-based assays as well as about interactions of TCC and MWCNT. Our results on the activity of TCC in the ER-responsive cells provide an explanation for the mechanism how chemicals enhance the endocrine-disrupting activity of chemicals [54]. Chemicals acting as endocrine-disrupting compounds (EDC) affect the ER receptor and lead to activation/inhibition of hormone-dependent gene expression[54]. However, EDC may also alter hormone receptor function simply by changing phosphorylation of the receptor (activating him) without the responsible chemical or natural ligand ever binding to the receptor [135].

Clearly, further examinations are required especially the confirmation of our findings in vivo.

Triclocarban at concentrations up to 1.6 µM showed no generation of ROS in three cell lines. Two similar studies suggested the production of reactive oxygen species in rat thymocytes after an incubation time of 1 h to 300 nM or higher concentrations of TCC [126,129]. On the contrary, Fukunaga and coworkers [128] supposed that the same cells recovered the initial loss of cellular glutathione as a biomarker of oxidative stress in the continued presence of 300 nM TCC. Thus, the ability of TCC to generate ROS in human cell lines is still under discussion and further research is required.

Interaction of MWCNT and TCC

Most reported studies have illustrated that the CNT surface area is an adsorbent for organic chemicals, such as polycyclic aromatic hydrocarbons, phenolic compounds, and endocrine disrupting chemicals [29,136,137]. In the present study, we determined for the first time lower cell toxicity in MWCNT- and TCC-treated H295R cells compared to the cytotoxic potential of TCC alone. Even the antiestrogenic potential of TCC in the ER Calux assay with T47Dluc cells was reduced in the presence of MWCNT compared to the absence of the nanotubes in the whole experimental design. To our knowledge, the influence of MWCNT on the availability of TCC was not examined before. The antimicrobial agent TCC seems to interact with MWCNT resulting in a lower available concentration of TCC in the test medium. This could be proven in the ER Calux assay (Figure 4). Treatment of the cells with higher levels of CNT combined with a lower TCC concentration (0.5% of the nanotubes) did not result in a decrease of luciferase activity compared to same concentrations of the antimicrobial biocide and the mixture of MWCNT and TCC (concentration 1% of that of CNT).

Only few studies have been conducted to understand the adsorption of organic contaminants by CNT [25-27,29,138-140]. A common observation from these studies was that CNT are very strong adsorbents for hydrophobic organic compounds. Possible adsorption mechanisms are the hydrophobic interactions between TCC and CNT or non-covalent ϖ-ϖ electron-donor-acceptor (EDA) interactions [141]. With a log K_{OW} of 4.9 for TCC [59] and considering the strong hydrophobicity and high surface area of carbon nanotubes [142], the hydrophobic effect might be the dominant factor for the adsorption of TCC on the MWCNT. Chen et al. [142] reported that the strong adsorption of polar nitroaromatics, compared to apolar compounds, was due to ϖ-ϖ EDA interactions between the nitroaromatics (ϖ acceptor) and the graphene sheets (ϖ donors) of CNT. An important implication from several of the studies is that electronic polarizability of the aromatic rings on the surface of CNT might considerably enhance adsorption of the organic compounds[25,138-140]. As concluded by Chen and coworkers [142], no studies have been conducted to systematically compare adsorptive interactions between carbon nanotubes and organic compounds with significantly different physical-chemical properties (e.g., polarity, functional groups, etc.). In addition, engineered carbon nanomaterials can vary significantly in shape, size and morphology, and impurity, e.g., metal, amorphous carbon, and O-containing groups, which can further complicate the adsorptive properties of these materials for organic contaminants [142].

CONCLUSIONS

We investigated the cytotoxicity and the endocrine potential of unfunctionalized, flexible MWCNT and their capability to enhance the production of intracellular ROS. TEM analyses revealed the presence of well-dispersed, isolated nanotubes as well as aggregated clusters in our assays. We found that the tested CNT are not toxic to RTL-W1, T47Dluc, and H295R cells. As assumed, we did not find a significant change in luciferase activity in the ER Calux assay with T47Dluc cells nor a significant alteration of E2 production in H295R cells after treatment with MWCNT. Consistent with other studies, this work also shows the generation of ROS by MWCNT.

Concentrations (1.6 μM) of the biocide TCC decreased the luciferase activity in ER Calux assays but did not affect the production of E2 in H295R cells in ELISA assays. In mixtures of MWCNT and TCC, the antiestrogen potential of TCC in T47Dluc cells was reduced because the lipophilic biocide adsorbed to the nanotubes resulting in a lower available concentration of TCC in the test medium. More research is needed to better understand the molecular interactions of carbon nanotubes and organic contaminants. In such experiments, the properties of both contaminants, CNT, and pollutants, should be systematically varied.

AUTHOR'S CONTRIBUTIONS

AS (first author) carried out the experimental studies and drafted the manuscript. SM enabled to carry out the in vitro testing of T47Dluc cells and helped to perform one part of the statistical analysis. HH conceived of the study and participated in its design. AS conceived of the study and participated in the sequence alignment. HM participated in the design of the study and helped to perform the statistical analysis and to draft the manuscript. All authors read and approved the final manuscript.

ACKNOWLEDGMENTS

We thank Simone Hotz from the Institute for Environmental Research at the RWTH Aachen University for supporting the practical work. The authors also thank the German Federal Ministry of Education and Research (BMBF) for funding the CarboLifeCycle project as a part of Inno.CNT, the innovation alliance for CNT (http://www.inno-cnt.de/en/). The authors would like to express their thanks to Drs. Niels C. Bols and Lucy Lee (University of Waterloo, Canada) for providing RTL-W1 cells and BioDetection Systems for the ER-CALUX cells.

REFERENCES

1. Ball P: Roll up for the revolution. Nature 2001, 414:142-144.
2. Dalton AB, Collins S, Munoz E, Razal JM, Ebron VH, Ferraris JP, Coleman JN, Kim BG, Baughman RH: Super-tough carbon-nanotube fibres. Nature 2003, 423:703-703.
3. Mauter MS, Elimelech M: Environmental applications of carbon-based nanomaterials. Environ Sci Technol 2008, 42:5843-5859.
4. Petersen EJ, Henry TB: Methodological considerations for testing the ecotoxicity of carbon nanotubes and fullerenes: review. Environ Toxicol Chem 2012, 31:60-72.
5. Haniu H, Saito N, Matsuda Y, Kim YA, Park KC, Tsukahara T, Usui Y, Aoki K, Shimizu M, Ogihara N, Hara K, Takanashi S, Okamoto M, Ishigaki N, Nakamura K, Kato H: Elucidation mechanism of different biological responses to multi-walled carbon nanotubes using four cell lines. Int J Nanomedicine 2011, 6:3487-3497.
6. Armstrong D, Bharali DJ, Armstrong D, Bharali DJ: Nanoparticles: toxicity, radicals, electron transfer, and antioxidants. In Oxidative Stress and Nanotechnology. Northern Algeria: Human Press; 2013:16-17.
7. EU - European Commission Recommendation on the definition of nanomaterialhttp://ec.europa.eu/environment/chemicals/nanotech/faq/definition_en.htmwebcite
8. Thurnherr T, Brandenberger C, Fischer K, Diener L, Manser P, Maeder-Althaus X, Kaiser J-P, Krug HF, Rothen-Rutishauser B, Wick P: A comparison of acute and long-term effects of industrial multiwalled carbon nanotubes on human lung and immune cells in vitro. Toxicol Lett 2011, 200:176-186.
9. Rotoli BM, Bussolati O, Bianchi MG, Barilli A, Balasubramanian C, Bellucci S, Bergamaschi E:Non-functionalized multi-walled carbon nanotubes alter the paracellular permeability of human airway epithelial cells. Toxicol Lett 2008, 178:95-102.

10. Foley S, Crowley C, Smaihi M, Bonfils C, Erlanger BF, Seta P, Larroque C: Cellular localisation of a water-soluble fullerene derivative. Biochem Biophys Res Commun 2002, 294:116-119.
11. Lu Q, Moore JM, Huang G, Mount AS, Rao AM, Larcom LL, Ke PC: RNA polymer translocation with single-walled carbon nanotubes. Nano Lett 2004, 4:2473-2477.
12. Shi Kam NW, Jessop TC, Wender PA, Dai H: Nanotube molecular transporters: internalization of carbon nanotube-protein conjugates into mammalian cells. J Am Chem Soc 2004, 126:6850-6851.
13. Schinwald A, Donaldson K: Use of back-scatter electron signals to visualise cell/nanowires interactions in vitro and in vivo; frustrated phagocytosis of long fibres in macrophages and compartmentalisation in mesothelial cells in vivo. Part Fibre Toxicol 2012, 9:34.
14. Shvedova AA, Kisin ER, Mercer R, Murray AR, Johnson VJ, Potapovich AI, Tyurina YY, Gorelik O, Arepalli S, Schwegler-Berry D: Unusual inflammatory and fibrogenic pulmonary responses to single-walled carbon nanotubes in mice. AJP Lung 2005, 289:L698-L708.
15. Stellaa GM: Carbon nanotubes and pleural damage: perspectives of nanosafety in the light of asbestos experience. Biointerphases 2011, 6:P1-P17.
16. Cui D, Tian F, Ozkan CS, Wang M, Gao H: Effect of single wall carbon nanotubes on human HEK293 cells. Toxicol Lett 2005, 155:73-85.
17. Jia G, Wang H, Yan L, Wang X, Pei R, Yan T, Zhao Y, Guo X: Cytotoxicity of carbon nanomaterials: single-wall nanotube, multi-wall nanotube, and fullerene. Environ Sci Technol 2005, 39:1378-1383.
18. Monteiro-Riviere NA, Nemanich RJ, Inman AO, Wang YY, Riviere JE: Multi-walled carbon nanotube interactions with human epidermal keratinocytes. Toxicol Lett 2005, 155:377-384.

19. Shvedova A, Castranova V, Kisin E, Schwegler-Berry D, Murray A, Gandelsman V, Maynard A, Baron P: Exposure to carbon nanotube material: assessment of nanotube cytotoxicity using human keratinocyte cells. J Toxicol Environ Health A 2003, 66:1909-1926.
20. Warheit DB, Laurence B, Reed KL, Roach D, Reynolds G, Webb T: Comparative pulmonary toxicity assessment of single-wall carbon nanotubes in rats. Toxicol Sci 2004, 77:117-125.
21. Borm PJ: Particle toxicology: from coal mining to nanotechnology. Inhalation Toxicol 2002, 14:311-324.
22. Brumfiel G: Nanotechnology: a little knowledge. Nature 2003, 424:246-248.
23. Colvin VL: The potential environmental impact of engineered nanomaterials. Nat Biotechnol 2003, 21:1166-1170.
24. Gotovac S, Yang C-M, Hattori Y, Takahashi K, Kanoh H, Kaneko K: Adsorption of polyaromatic hydrocarbons on single wall carbon nanotubes of different functionalities and diameters. J Colloid Interface Sci 2007, 314:18-24.
25. Long RQ, Yang RT: Carbon nanotubes as superior sorbent for dioxin removal. J Am Chem Soc 2001, 123:2058-2059.
26. Lu C, Chung Y-L, Chang K-F: Adsorption thermodynamic and kinetic studies of trihalomethanes on multiwalled carbon nanotubes. J Hazard Mater 2006, 138:304-310.
27. Peng X, Li Y, Luan Z, Di Z, Wang H, Tian B, Jia Z: Adsorption of 1,2-dichlorobenzene from water to carbon nanotubes. Chem Phys Lett 2003, 376:154-158.
28. Upadhyayula VK, Deng S, Mitchell MC, Smith GB: Application of carbon nanotube technology for removal of contaminants in drinking water: a review. Sci Total Environ 2009, 408:1-13.
29. Yang K, Zhu L, Xing B: Adsorption of polycyclic aromatic hydrocarbons by carbon nanomaterials. Environ Sci Technol 2006, 40:1855-1861.
30. Zhang S, Shao T, Kose HS, Karanfil T: Adsorption of aromatic compounds by carbonaceous adsorbents: a comparative

study on granular activated carbon, activated carbon fiber, and carbon nanotubes. Environ Sci Technol 2010, 44:6377-6383.
31. Zhang S, Shao T, Kose HS, Karanfil T: Adsorption kinetics of aromatic compounds on carbon nanotubes and activated carbons. Environ Toxicol Chem 2012, 31:79-85.
32. Savage N, Diallo MS: Nanomaterials and water purification: opportunities and challenges. J Nanopart Res 2005, 7:331-342.
33. Di Z-C, Ding J, Peng X-J, Li Y-H, Luan Z-K, Liang J: Chromium adsorption by aligned carbon nanotubes supported ceria nanoparticles. Chemosphere 2006, 62:861-865.
34. Li Y-H, Di Z, Ding J, Wu D, Luan Z, Zhu Y: Adsorption thermodynamic, kinetic and desorption studies of Pb^{2+} on carbon nanotubes. Water Res 2005, 39:605-609.
35. Rao GP, Lu C, Su F: Sorption of divalent metal ions from aqueous solution by carbon nanotubes: a review. Sep Purif Technol 2007, 58:224-231.
36. Peng X, Luan Z, Ding J, Di Z, Li Y, Tian B: Ceria nanoparticles supported on carbon nanotubes for the removal of arsenate from water. Mater Lett 2005, 59:399-403.
37. Yan X, Shi B, Lu J, Feng C, Wang D, Tang H: Adsorption and desorption of atrazine on carbon nanotubes. J Colloid Interface Sci 2008, 321:30-38.
38. Akasaka T, Watari F: Capture of bacteria by flexible carbon nanotubes. Acta Biomater 2009, 5:607-612.
39. Deng J, Yu L, Liu C, Yu K, Shi X, Yeung LWY, Lam PKS, Wu RSS, Zhou B:Hexabromocyclododecane-induced developmental toxicity and apoptosis in zebrafish embryos. Aquat Toxicol 2009, 93:29-36.
40. Upadhyayula VK, Deng S, Smith GB, Mitchell MC: Adsorption of Bacillus subtilis on single-walled carbon nanotube aggregates, activated carbon and NanoCeram™. Water Res 2009, 43:148-156.

41. BradyEstévez AS, Kang S, Elimelech M: A singlewalledcarbon nanotube filter for removal of viral and bacterial pathogens. Small 2008, 4:481-484.
42. Mostafavi S, Mehrnia M, Rashidi A: Preparation of nanofilter from carbon nanotubes for application in virus removal from water. Desalination 2009, 238:271-280.
43. Albuquerque Júnior EC, Méndez MOA, Coutinho AR, Franco TT: Removal of cyanobacteria toxins from drinking water by adsorption on activated carbon fibers. Mater Res 2008, 11:371-380.
44. Yan H, Gong A, He H, Zhou J, Wei Y, Lv L: Adsorption of microcystins by carbon nanotubes. Chemosphere 2006, 62:142-148.
45. Hyung H, Kim J-H: Natural organic matter (NOM) adsorption to multi-walled carbon nanotubes: effect of NOM characteristics and water quality parameters. Environ Sci Technol 2008, 42:4416-4421.
46. Lu C, Su F: Adsorption of natural organic matter by carbon nanotubes. Sep Purif Technol 2007, 58:113-121.
47. Saleh NB, Pfefferle LD, Elimelech M: Aggregation kinetics of multiwalled carbon nanotubes in aquatic systems: measurements and environmental implications. Environ Sci Technol 2008, 42:7963-7969.
48. Bottini M, Bruckner S, Nika K, Bottini N, Bellucci S, Magrini A, Bergamaschi A, Mustelin T:Multi-walled carbon nanotubes induce T lymphocyte apoptosis. Toxicol Lett 2006, 160:121-126.
49. Ding L, Stilwell J, Zhang T, Elboudwarej O, Jiang H, Selegue JP, Cooke PA, Gray JW, Chen FF: Molecular characterization of the cytotoxic mechanism of multiwall carbon nanotubes and nano-onions on human skin fibroblast. Nano Lett 2005, 5:2448-2464.
50. Pulskamp K, Diabaté S, Krug HF: Carbon nanotubes show no sign of acute toxicity but induce intracellular reactive oxygen species in dependence on contaminants. Toxicol Lett 2007,

168:58-74.
51. Simon-Deckers A, Gouget B, Mayne-L'Hermite M, Herlin-Boime N, Reynaud C, Carriere M: In vitro investigation of oxide nanoparticle and carbon nanotube toxicity and intracellular accumulation in A549 human pneumocytes. Toxicology 2008, 253:137-146.
52. Klaine SJ, Alvarez PJJ, Batley GE, Fernandes TF, Handy RD, Lyon DY, Mahendra S, McLaughlin MJ, Lead JR: Nanomaterials in the environment: behavior, fate, bioavailability, and effects. Environ Toxicol Chem 2008, 27:1825-1851.
53. Brausch JM, Rand GM: A review of personal care products in the aquatic environment: environmental concentrations and toxicity. Chemosphere 2011, 82:1518-1532.
54. Ahn KC, Zhao B, Chen J, Cherednichenko G, Sanmarti E, Denison MS, Lasley B, Pessah IN, Kültz D, Chang DPY: In vitro biologic activities of the antimicrobials triclocarban, its analogs, and triclosan in bioassay screens: receptor-based bioassay screens. Environ Health Perspect 2008, 116:1203.
55. Agyin-Birikorang S, Miller M, O'Connor GA: Retention-release characteristics of triclocarban and triclosan in biosolids, soils, and biosolids-amended soils. Environ Toxicol Chem 2010, 29:1925-1933.
56. Hamilton W: Membrane-active antibacterial compounds. Biochem J 1970, 118:46P-47P.
57. TCC Consortium: High Production Volume (HPV) Chemical Challenge Program Data Availability and Screening Level Assessment for Triclocarban. CAS#: 101-20-2 2002 Report No 201-14186A; 2002. http://www.epa.gov/hpv/pubs/summaries/tricloca/c14186tc.htm webcite
58. Heidler J, Sapkota A, Halden RU: Partitioning, persistence, and accumulation in digested sludge of the topical antiseptic triclocarban during wastewater treatment. Environ Sci Technol 2006, 40:3634-3639.
59. Ying G-G, Yu X-Y, Kookana RS: Biological degradation of triclocarban and triclosan in a soil under aerobic and

anaerobic conditions and comparison with environmental fate modelling. Environ Pollut 2007, 150:300-305.

60. Chalew TE, Halden RU: Environmental exposure of aquatic and terrestrial biota to triclosan and triclocarban1. J Am Water Resour As 2009, 45:4-13.

61. Clarke BO, Smith SR: Review of 'emerging' organic contaminants in biosolids and assessment of international research priorities for the agricultural use of biosolids. Environ Int 2011, 37:226-247.

62. Miller TR, Colquhoun DR, Halden RU: Identification of wastewater bacteria involved in the degradation of triclocarban and its non-chlorinated congener. J Hazard Mater 2010, 183:766-772.

63. Kolpin DW, Furlong ET, Kolpin DW, Furlong ET, Meyer MT, Thurman EM, Zaugg SD, Barber LB, Buxton HT: Pharmaceuticals, hormones, and other organic wastewater contaminants in US streams, 1999–2000: a national reconnaissance. Environ Sci Technol 2002, 36:1202-1211.

64. Halden RU, Paull DH: Co-occurrence of triclocarban and triclosan in US water resources. Environ Sci Technol 2005, 39:1420-1426.

65. Coogan MA, Edziyie RE, La Point TW, Venables BJ: Algal bioaccumulation of triclocarban, triclosan, and methyl-triclosan in a North Texas wastewater treatment plant receiving stream. Chemosphere 2007, 67:1911-1918.

66. Darbre P: Environmental oestrogens, cosmetics and breast cancer. Best Pract Res Cl En 2006, 20:121-143.

67. Chen J, Ahn KC, Gee NA, Ahmed MI, Duleba AJ, Zhao L, Gee SJ, Hammock BD, Lasley BL:Triclocarban enhances testosterone action: a new type of endocrine disruptor? Endocrinology 2008, 149:1173-1179.

68. Hollert H, Dürr M, Erdinger L, Braunbeck T: Cytotoxicity of settling particulate matter (SPM) and sediments of the Neckar river (Germany) during a winter flood. Environ Toxicol Chem 2000, 19:528-534.

69. Arechabala B, Coiffard C, Rivalland P, Coiffard L, Roeck-Holtzhauer YD: Comparison of cytotoxicity of various surfactants tested on normal human fibroblast cultures using the neutral red test, MTT assay and LDH release. J Appl Toxicol 1999, 19:163-165.
70. Borenfreund E, Babich H, Martin-Alguacil N: Comparisons of two in vitro cytotoxicity assays—the neutral red (NR) and tetrazolium MTT tests. Toxicol In Vitro 1988, 2:1-6.
71. Fotakis G, Timbrell JA: In vitro cytotoxicity assays: comparison of LDH, neutral red, MTT and protein assay in hepatoma cell lines following exposure to cadmium chloride. Toxicol Lett 2006, 160:171-177.
72. Legler J, van den Brink CE, Brouwer A, Murk AJ, van der Saag PT, Vethaak AD, van der Burg B: Development of a stably transfected estrogen receptor-mediated luciferase reporter gene assay in the human T47D breast cancer cell line. Toxicol Sci 1999, 48:55.
73. Hecker M, Hollert H, Cooper R, Vinggaard AM, Akahori Y, Murphy M, Nellemann C, Higley E, Newsted J, Laskey J: The OECD validation program of the H295R steroidogenesis assay: phase 3. Final inter-laboratory validation study. Environ Sci Pollut R 2011, 18:503-515.
74. Hecker M, Hollert H, Cooper R, Vinggaard A-M, Akahori Y, Murphy M, Nellemann C, Higley E, Newsted J, Wu R: The OECD validation program of the H295R steroidogenesis assay for the identification of in vitro inhibitors and inducers of testosterone and estradiol production. Phase 2: inter-laboratory pre-validation studies. Environ Sci Pollut R 2007, 14:23-30.
75. Gracia T, Jones PD, Higley EB, Hilscherova K, Newsted JL, Murphy MB, Chan AK, Zhang X, Hecker M, Lam PK: Modulation of steroidogenesis by coastal waters and sewage effluents of Hong Kong, China, using the H295R assay. Environ Sci Pollut R Int 2008, 15:332-343.
76. Hecker M, Hollert H: Effect-directed analysis (EDA) in aquatic

ecotoxicology: state of the art and future challenges. Environ Sci Pollut R 2009, 16:607-613.
77. Kase R, Kunz P, Gerhardt A: Identifikation geeigneter Nachweismöglichkeiten von hormonaktiven und reproduktionstoxischen Wirkungen in aquatischen Ökosystemen. Umweltwiss Schadstoff-Forsch 2009, 21:339-378.
78. Leusch FD, De Jager C, Levi Y, Lim R, Puijker L, Sacher F, Tremblay LA, Wilson VS, Chapman HF: Comparison of five in vitro bioassays to measure estrogenic activity in environmental waters. Environ Sci Technol 2010, 44:3853-3860.
79. Hecker M, Hollert H: Endocrine disruptor screening: regulatory perspectives and needs. ESEU 2011, 23:1-14.
80. Grund S, Higley E, Schönenberger R, Suter MJ, Giesy JP, Braunbeck T, Hecker M, Hollert H:The endocrine disrupting potential of sediments from the Upper Danube River (Germany) as revealed by in vitro bioassays and chemical analysis. Environ Sci Pollut R 2011, 18:446-460.
81. LeBel CP, Ischiropoulos H, Bondy SC: Evaluation of the probe 2',7'-dichlorofluorescin as an indicator of reactive oxygen species formation and oxidative stress. Chem Res Toxicol 1992, 5:227-231.
82. Lee LE, Clemons JH, Bechtel DG, Caldwell SJ, Han K-B, Pasitschniak-Arts M, Mosser DD, Bols NC: Development and characterization of a rainbow trout liver cell line expressing cytochrome P450-dependent monooxygenase activity. Cell Biol Toxicol 1993, 9:279-294.
83. Klee N, Gustavsson L, Kosmehl T, Engwall M, Erdinger L, Braunbeck T, Hollert H: Changes in toxicity and genotoxicity of industrial sewage sludge samples containing nitro- and amino-aromatic compounds following treatment in bioreactors with different oxygen regimes. Environ Sci Pollut R 2004, 11:313-320.
84. Maletz S, Floehr T, Beier S, Klumper C, Brouwer A, Behnisch P, Higley E, Giesy JP, Hecker M, Gebhardt W, Linnemann

V, Pinnekamp J, Hollert H: In vitro characterization of the effectiveness of enhanced sewage treatment processes to eliminate endocrine activity of hospital effluents. Water Res 2013, 47:1545-1557.

85. Hilscherova K, Jones PD, Gracia T, Newsted JL, Zhang X, Sanderson J, Richard M, Wu RS, Giesy JP: Assessment of the effects of chemicals on the expression of ten steroidogenic genes in the H295R cell line using real-time PCR. Toxicol Sci 2004, 81:78-89.

86. Borenfreund E, Puerner JA: A simple quantitative procedure using monolayer cultures for cytotoxicity assays (HTD/NR-90). J Tissue Cult Methods 1985, 9:7-9.

87. Heger S, Bluhm K, Agler MT, Maletz S, Schäffer A, Seiler T-B, Angenent LT, Hollert H:Biotests for hazard assessment of biofuel fermentation. Energ Environ Sci 2012, 5:9778-9788.

88. Wang JY, Sun PP, Bao YM, Liu JW, An LJ: Cytotoxicity of single-walled carbon nanotubes on PC12 cells. Toxicol In Vitro 2011, 25:242-250.

89. Blaha L, Hecker M, Murphy M, Jones P, Giesy JP: Procedure for determination of cell viability/cytotoxicity using the MTT bioassay. Michigan: Aquatic Toxicology Laboratory, Michigan State University; 2004.

90. Mosmann T: Rapid colorimetric assay for cellular growth and survival: application to proliferation and cytotoxicity assays. J Immunol Methods 1983, 65:55-63.

91. Houtman CJ, Cenijn PH, Hamers T, Lamoree MH, Legler J, Murk AJ, Brouwer A:Toxicological profiling of sediments using in vitro bioassays, with emphasis on endocrine disruption. Environ Toxicol Chem 2004, 23:32-40.

92. Barillet S, Simon-Deckers A, Herlin-Boime N, Mayne-L'Hermite M, Reynaud C, Cassio D, Gouget B, Carriere M: Toxicological consequences of TiO2, SiC nanoparticles and multi-walled carbon nanotubes exposure in several mammalian cell types: an in vitro study. J Nanopart Res 2010, 12:61-73.

93. Jacobsen NR, Pojana G, White P, Moller P, Cohn CA, Korsholm KS, Vogel U, Marcomini A, Loft S, Wallin H: Genotoxicity, cytotoxicity, and reactive oxygen species induced by single-walled carbon nanotubes and C-60 fullerenes in the FE1-Muta (TM) mouse lung epithelial cells. Environ Mol Mutag 2008, 49:476-487.

94. Pietsch C, Bucheli TD, Wettstein FE, Burkhardt-Holm P: Frequent biphasic cellular responses of permanent fish cell cultures to deoxynivalenol (DON). Toxicol Appl Pharmacol 2011, 256:24-34.

95. Sohaebuddin SK, Thevenot PT, Baker D, Eaton JW, Tang LP: Nanomaterial cytotoxicity is composition, size, and cell type dependent. Part Fibre Toxicol 2010, 7:22.

96. Shukla A, Ramos-Nino M, Mossman B: Cell signaling and transcription factor activation by asbestos in lung injury and disease. Int J Biochem Cell 2003, 35:1198-1209.

97. Di Giorgio ML, Bucchianico SD, Ragnelli AM, Aimola P, Santucci S, Poma A: Effects of single and multi walled carbon nanotubes on macrophages: cyto and genotoxicity and electron microscopy. Mutat Res-Gen Tox En 2011, 722:20-31.

98. Tian F, Cui D, Schwarz H, Estrada GG, Kobayashi H: Cytotoxicity of single-wall carbon nanotubes on human fibroblasts. Toxicol In Vitro 2006, 20:1202-1212.

99. Donaldson K, Poland CA: Nanotoxicity: challenging the myth of nano-specific toxicity. Curr Opin Biotechnol 2013, 24:724-734.

100. Coccini T, Roda E, Sarigiannis DA, Mustarelli P, Quartarone E, Profumo A, Manzo L: Effects of water-soluble functionalized multi-walled carbon nanotubes examined by different cytotoxicity methods in human astrocyte D384 and lung A549 cells. Toxicology 2010, 269:41-53.

101. Magrez A, Kasas S, Salicio V, Pasquier N, Seo JW, Celio M, Catsicas S, Schwaller B, Forró L:Cellular toxicity of carbon-based nanomaterials. Nano Lett 2006, 6:1121-1125.

102. Ye S-F, Wu Y-H, Hou Z-Q, Zhang Q-Q: ROS and NF- B are involved in upregulation of IL-8 in A549 cells exposed to multi-walled carbon nanotubes. Biochem Biophys Res Commun 2009, 379:643-648.
103. Hu XK, Cook S, Wang P, Hwang HM, Liu X, Williams QL: In vitro evaluation of cytotoxicity of engineered carbon nanotubes in selected human cell lines. Sci Total Environ 2010, 408:1812-1817.
104. Kisin ER, Murray AR, Keane MJ, Shi X-C, Schwegler-Berry D, Gorelik O, Arepalli S, Castranova V, Wallace WE, Kagan VE: Single-walled carbon nanotubes: geno- and cytotoxic effects in lung fibroblast V79 cells. J Toxicol Environ Health A 2007, 70:2071-2079.
105. Pacurari M, Yin XJ, Zhao J, Ding M, Leonard SS, Schwegler-Berry D, Ducatman BS, Sbarra D, Hoover MD, Castranova V: Raw single-wall carbon nanotubes induce oxidative stress and activate MAPKs, AP-1, NF- B, and Akt in normal and malignant human mesothelial cells. Environ Health Perspect 2008, 116:1211.
106. Lindberg HK, Falck GC-M, Suhonen S, Vippola M, Vanhala E, Catalán J, Savolainen K, Norppa H: Genotoxicity of nanomaterials: DNA damage and micronuclei induced by carbon nanotubes and graphite nanofibres in human bronchial epithelial cells in vitro. Toxicol Lett 2009, 186:166-173.
107. Belyanskaya L, Manser P, Spohn P, Bruinink A, Wick P: The reliability and limits of the MTT reduction assay for carbon nanotubes–cell interaction.
108. Carbon 2007, 45:2643-2648.
109. Davoren M, Herzog E, Casey A, Cottineau B, Chambers G, Byrne HJ, Lyng FM: In vitro toxicity evaluation of single walled carbon nanotubes on human A549 lung cells. Toxicol In Vitro 2007, 21:438-448.
110. Warheit DB: How meaningful are the results of nanotoxicity studies in the absence of adequate material characterization?

Toxicol Sci 2008, 101:183-185.

111. Aschberger K, Johnston HJ, Stone V, Aitken RJ, Hankin SM, Peters SAK, Tran CL, Christensen FM: Review of carbon nanotubes toxicity and exposure - appraisal of human health risk assessment based on open literature. Crit Rev Toxicol 2010, 40:759-790.

112. Crouzier D, Follot S, Gentilhomme E, Flahaut E, Arnaud R, Dabouis V, Castellarin C, Debouzy JC: Carbon nanotubes induce inflammation but decrease the production of reactive oxygen species in lung. Toxicology 2010, 272:39-45.

113. Yang ST, Wang X, Jia G, Gu YQ, Wang TC, Nie HY, Ge CC, Wang HF, Liu YF: Long-term accumulation and low toxicity of single-walled carbon nanotubes in intravenously exposed mice. Toxicol Lett 2008, 181:182-189.

114. Murr LE, Garza KM, Soto KF, Carrasco A, Powell TG, Ramirez DA, Guerrero PA, Lopez DA, Venzor J 3rd: Cytotoxicity assessment of some carbon nanotubes and related carbon nanoparticle aggregates and the implications for anthropogenic carbon nanotube aggregates in the environment. Int J Env Res Public Health 2005, 2:31-42.

115. Chen B, Liu Y, Song WM, Hayashi Y, Ding XC, Li WH: In vitro evaluation of cytotoxicity and oxidative stress induced by multiwalled carbon nanotubes in murine RAW 264.7 macrophages and human A549 lung cells. Biomed Environ Sci 2011, 24:593-601.

116. Pulskamp K, Wörle-Knirsch JM, Hennrich F, Kern K, Krug HF: Human lung epithelial cells show biphasic oxidative burst after single-walled carbon nanotube contact. Carbon 2007, 45:2241-2249.

117. Wörle-Knirsch J, Pulskamp K, Krug H: Oops they did it again! Carbon nanotubes hoax scientists in viability assays. Nano Lett 2006, 6:1261-1268.

118. Karlsson HL, Cronholm P, Gustafsson J, Moller L: Copper oxide nanoparticles are highly toxic: A comparison between metal oxide nanoparticles and carbon nanotubes. Chem Res

Toxicol 2008, 21:1726-1732.

119. Vittorio O, Raffa V, Cuschieri A: Influence of purity and surface oxidation on cytotoxicity of multiwalled carbon nanotubes with human neuroblastoma cells. Nanosci Nanotechnol Biol Med 2009, 5:424-431.

120. Xu H, Bai J, Meng J, Hao W, Xu H, Cao J-M: Multi-walled carbon nanotubes suppress potassium channel activities in PC12 cells. Nanotechnology 2009, 20:285102.

121. Ye S, Wang Y, Jiao F, Zhang H, Lin C, Wu Y, Zhang Q: The role of NADPH oxidase in multi-walled carbon nanotubes-induced oxidative stress and cytotoxicity in human macrophages. J Nanosci Nanotechnol 2011, 11:3773-3781.

122. Yang L, Ying L, Yujian F, Taotao W, Le Guyader L, Ge G, Ru-Shi L, Yan-Zhong C, Chunying C: The triggering of apoptosis in macrophages by pristine graphene through the MAPK and TGF-beta signaling pathways. Biomaterials 2012, 33:402-411.

123. Zhang Y, Ali SF, Dervishi E, Xu Y, Li Z, Casciano D, Biris AS: Cytotoxicity effects of graphene and single-wall carbon nanotubes in neural phaeochromocytoma-derived PC12 cells. ACS Nano 2010, 4:3181-3186.

124. Chang Y, Yang S-T, Liu J-H, Dong E, Wang Y, Cao A, Liu Y, Wang H: In vitro toxicity evaluation of graphene oxide on A549 cells. Toxicol Lett 2011, 200:201-210.

125. Creighton MA, Rangel-Mendez JR, Huang JX, Kane AB, Hurt RH: Graphene-Induced Adsorptive and Optical Artifacts During In Vitro Toxicology Assays. Small 2013, 9:1921-1927.

126. Lawrence J, Zhu B, Swerhone G, Roy J, Wassenaar L, Topp E, Korber D: Comparative microscale analysis of the effects of triclosan and triclocarban on the structure and function of river biofilm communities. Sci Total Environ 2009, 407:3307-3316.

127. Morita J, Teramachi A, Sanagawa Y, Toyson S, Yamamoto H, Oyama Y: Elevation of intracellular Zn^{2+} level by nanomolar concentrations of triclocarban in rat thymocytes. Toxicol Lett

2012, 215:208-2013.

128. Snyder EH, O'Connor GA, McAvoy DC: Toxicity and bioaccumulation of biosolids-borne triclocarban (TCC) in terrestrial organisms. Chemosphere 2010, 408:2667-2673.

129. Fukunaga E, Kanbara Y, Oyama Y: Role of Zn^{2+} in restoration of nonprotein thiol content in the cells under chemical stress induced by triclocarban. Nat Sci Res 2013, 27:1-5.

130. Kanbara Y, Murakane K, Nishimura Y, Satoh M, Oyama Y: Nanomolar concentration of triclocarban increases the vulnerability of rat thymocytes to oxidative stress. J Toxicol Sci 2013, 38:49-55.

131. Legler J, Zeinstra LM, Schuitemaker F, Lanser PH, Bogerd J, Brouwer A, Vethaak AD, De Voogt P, Murk AJ, van der Burg B: Comparison of in vivo and in vitro reporter gene assays for short-term screening of estrogenic activity. Environ Sci Technol 2002, 36:4410-4415.

132. Tarnow P, Tralau T, Hunecke D, Luch A: Effects of triclocarban on the transcription of estrogen, androgen and aryl hydrocarbon receptor responsive genes in human breast cancer cells. Toxicol In Vitro 2013, 27:1467-1475.

133. Thorne N, Auld DS, Inglese J: Apparent activity in high-throughput screening: origins of compound-dependent assay interference. Curr Opin Chem Biol 2010, 14:315-324.

134. Thorne N, Shen M, Lea WA, Simeonov A, Lovell S, Auld DS, Inglese J: Firefly luciferase in chemical biology: a compendium of inhibitors, mechanistic evaluation of chemotypes, and suggested use as a reporter. Chem Biol 2012, 19:1060-1072.

135. Sotoca A, Bovee T, Brand W, Velikova N, Boeren S, Murk A, Vervoort J, Rietjens I:Superinduction of estrogen receptor mediated gene expression in luciferase based reporter gene assays is mediated by a post-transcriptional mechanism. J Steroid Biochem Mol Biol 2010, 122:204-211.

136. Weigel NL, Moore NL: Steroid receptor phosphorylation: a key modulator of multiple receptor functions. Mol Endocrinol 2007, 21:2311-2319.

137. Lin D, Xing B: Adsorption of phenolic compounds by carbon nanotubes: role of aromaticity and substitution of hydroxyl groups. Environ Sci Technol 2008, 42:7254-7259.
138. Pan B, Lin D, Mashayekhi H, Xing B: Adsorption and hysteresis of bisphenol A and 17 alpha-ethinyl estradiol on carbon nanomaterials (vol 42, pg 5480, 2008). Environ Sci Technol 2009, 43:548-548.
139. Fagan SB, Souza Filho A, Lima J, Filho JM, Ferreira O, Mazali I, Alves O, Dresselhaus M: 1,2-Dichlorobenzene interacting with carbon nanotubes. Nano Lett 2004, 4:1285-1288.
140. Hilding J, Grulke EA, Sinnott SB, Qian D, Andrews R, Jagtoyen M: Sorption of butane on carbon multiwall nanotubes at room temperature. Langmuir 2001, 17:7540-7544.
141. Zhao J, Lu JP, Han J, Yang C-K: Noncovalent functionalization of carbon nanotubes by aromatic organic molecules. Appl Phys Lett 2003, 82:3746-3748.
142. Keiluweit M, Kleber M: Molecular-level interactions in soils and sediments: the role of aromatic ϖ-systems. Environ Sci Technol 2009, 43:3421-3429.
143. Chen W, Duan L, Zhu D: Adsorption of polar and nonpolar organic chemicals to carbon nanotubes. Environ Sci Technol 2007, 41:8295-8300.

Chapter 3

Optimal Parameters for in Vitro Development of the Fungus Hydrocarbonoclastic Penicillium sp.

Marcia Eugenia Ojeda-Morales[1], Miguel Ángel Hernández-Rivera[1], José Gabriel Martínez-Vázquez[2], Yolanda Córdova-Bautista[1], and Yuridia Evelin Hernández-Cardeño[1]

[1]División Académica de Ingeniería y Arquitectura, Universidad Juárez Autónoma de Tabasco, Cunduacán, México

[2]Facultad de Salud y Ciencias A. F., Universidad SEK, Santiago, Chile

ABSTRACT

México has extensive areas that have been impacted by oil spills for several decades. Current bioremediation technologies mostly used microorganisms to decontaminate sites with hydrocarbons. This research evaluated the conditions for the optimal development of the strain of a hydrocarbonoclastic fungus, which was found in samples of soil contaminated with 4.0×10^5 mg·kg^{-1} of Total Petroleum Hydrocarbons (TPH). A completely randomized experimental design with a $3 \times 3 \times 4$ factor arrangement was used: three levels of temperature ($T_1 = 29°C$, $T_2 = 35°C$ and $T_3 = 40°C$), three of pH ($pH_1 = 3.5$, $pH_2 = 5.0$ and $pH_3 = 6.0$) and four nutrients (N_1 = Urea, N_2 = Triple-17, N_3 = Nitrophoska-Blue and N_4 = Pure-Salts). Total fungi were isolated from the sampled soil and were sown in a combined carbon medium for hydrocarbonoclastic fungi and a strain was selected to be adapted to a liquid mineral medium. The selected strain was classified as Penicillium sp. Analyses of variance and mean tests were performed, using the SPSS-11.0 statistical software. The microorganisms showed the highest population growth in the treatment $N_2pH_2T_1$, which reached a value of 2.1×10^6 CFU·mL^{-1} in a biorreactor. To reach it, by bioaugmentation, the same development of Penicillium sp. in a conditioned soil would allow to implement a bioremediation strategy with great potential to retrieve soil contaminated with hydrocarbons both in Tabasco and in general in Mexico.

INTRODUCTION

The exploitation of energy resources in the state of Tabasco is of great importance for Mexico [1]. The total amount of crude oil extraction in 2009 in Mexico was 2.6 million of daily barrels. 77.5% of these barrels were obtained from the Mexican Gulf [2], and the other 22.5% were from wells in land. 77% of the petroleum extracted from the wells in land came from fields located in the state of Tabasco [3], which represented the 3.7% of the Gross National Product

(GNP) [4]. However, oil industry accidents like rupture of pipelines and shipwrecks cause oil spills on land and marine ecosystems, producing pollution processes that affect the biophysical properties of soil and water and the biological components of these ecosystems. All this prevents the polluted natural resources from being fully exploited as well as health problems to the population living there [5, 6]. Several studies including bioremediation techniques have been carried out in these sites, to stop the problems affecting the ecology and human health.

Bioremediation uses the metabolic potential of microorganism to clean up an open polluted environment [7]. In the oil industry, it is becoming a widely used and cost effective technique to clean up hydrocarbons due to its simple use over large areas and its capacity to completely destroy the contaminant [8, 9]. Within bioremediation, the bioaugmentation involves the inoculation of a strain or an enriched mixed microbial consortium in the soil [10-16]. It was found that indigenous microorganisms are especially efficient in the degradation of indigenous crude oil (produced in a determined region), but they are not that efficient in oils coming from other sites [17]. Native or indigenous microbes are present in small quantities and cannot prevent the contaminant from being spread; they do not have the capacity to degrade a particular contaminant or they can be in an inactive metabolic form in their habitats, which is why bioaugmentation is preferred over biostimulation [18]. Bioaugmentation offers an option to grow specific microbes in sufficient amounts to complete the biodegradation [15,16,19]. In this context, there exist a great number of heterotrophic microorganisms capable of using oil hydrocarbons as a source of carbon and energy, producing carbon dioxide, water, biomass and other less toxic products [20]. Among the group of fungi, the most extensive studies have focused on white rot fungi [21,22]. Filamentous fungi have some features that make them excellent agents of degradation. Once the microorganisms quickly ramify and the fungal hyphae penetrate the polluted soil, they reach the substratum and absorb it through the secretion of extracellular enzymes. They grow under stress conditions: under pH, lack of nutrients and low water activity [23]. Fungi have also demonstrated their ability to degrade in some cases, such as mineralize phenols, halogenated phenolic compounds, petroleum hydrocarbons, polycyclic aromatic hydrocarbons and polychlorinated biphenyls in large stress conditions [24,25].

The Penicillium sp. fungi have been used in studies of crude oil biodegradation [16,23,25-28], and the mineralization of a variety of oil derivates, such as polycyclic aromatic hydrocarbons (PAHs), including pyrene, chrysene, and benzo [a] pyrene and polar metabolites [12,25, 29-31], as well as toxicity studies with hexadecane, phenanthrene and beta-naphthol, as biodegradation sometimes cannot be carried out because of the toxicity of the oil rather than its own persistence [9]. The metabolism of PAHs by strains of filamentous fungi is mediated by extracellular lignolytic enzymes or intracellular cytochrome P450 monooxygenases [32,33]. Both routes mainly produce quinones as oxidation products, which have higher solubility and reactivity than starting PAH [34-36]. In the case of the Penicillium fungi, the monooxygenase enzyme systems are responsible for degrading PAHs, where the first steps of oxidation are the formation of monophenols, diphenols, dihydrodiol and quinone [35]. In a second step, conjugates of O-methyl and sulfate can be formed; they are detoxification products soluble in water [37].

It has been reported that fungi are better than bacteria in the degradation of hydrocarbons in both quantity and variety of constituents, indicating that the species of fungi most commonly found in soil and seawater contaminated with hydrocarbons are the genera Aspergillus and Penicillium [9,16,38,39]. The ability of Aspergillus and Penicillium sp. to tolerate these pollutants and grow in them, suggests they may be used as bioremediation agents and can be used in the restoration of the ecosystem when impacted by these pollutants, because they produce enzymes and acids that break and dismantle the long chains of hydrocarbons, the base structure common to oils, petroleum products and many other pollutants [29]. Despite the abundance of fungi, Penicillium in particular has received little attention in studies of biodegradation, with a relatively small number of pilot projects and large-scale studies using this organism that has a high potential in the field of bioremediation [25].

The objective of this study was to evaluate the optimal physicochemical and nutrient conditions that promote higher biomass growth of a hydrocarbonoclastic strain of the Penicillium sp., in order to have new strategies that allow the use of this strain in bioremediation by bioaugmentation in soil contaminated with crude oil or petroleum products, both in the state of Tabasco and throughout Mexico.

MATERIAL AND METHODS

The research was conducted in two stages: the first consisted of three phases where the process of isolation, purification, testing, characterization, and preservation of hydrocarbonoclastic microorganisms was developed.

Stage 1. Assessment, Characterization and Adaptation of Hydrocarbonoclastic Fungi.

Samples from a Gleysol-mollic contaminated soil were collected in the facilities of an oil field located in the Huimanguillo municipality [40], state of Tabasco in Mexico, at an altitude of 20 meters above sea level [41]. Simple soil samples were taken, in accordance with the Mexican Official Standard 021 [42], for a total of 5000 m^2 (0.5 ha), and stored at 4.0°C until use. Then, the Total Petroleum Hydrocarbons (TPH) present in the soil samples were determined by the method of extraction [43,44], in their heavy fraction (greater than C18) and then quantified by gravimetry [45]. Soil samples were subjected to sowing in Phase I.

Phase I

The process of isolation, purification and preservation of total fungi was based on their growth and reproduction in the Potato

Dextrose Agar growth medium PDA (Baker). Prepared according to the manufacturer, it was poured in petri dishes and sterilized. A solution was prepared with 10.0 g of the sampled soil in 90 mL of sterile distilled water, from which two serial dilutions were prepared: 1 mL of this solution was added to 9 mL of sterile distilled water (10^{-1}), making and additional dilution, similar to the latter solution (10^{-2}). Subsequently, 0.1 mL of the 10^{-1} dilution was sown in each of 6 the petri dishes with PDA, spreading with a Drigalsky spatula. The same was done with the 10^{-2} dilution and all of them were incubated at 28°C for 8 days (d). The quantification of total fungi was performed by the plate counts method for viable cells by serial dilution [46]. Selection tests of fungi in the soil were based on radial growth in the growth medium. Among the 24 strains of fungi found, 5 with the highest development were chosen to be assessed as hydrocarbonoclastic in Phase II (H14, H18, H19, H20 and H24). These strains were preserved on PDA medium in inclined tube at 4°C until use.

Phase II

Adaptation in Solid Medium Cellulose-Agar (SCA) for fungi to degrade total petroleum hydrocarbons. The SCA medium (Baker) was prepared according to Rivera-Cruz et al. [47], and poured into petri dishes. Culture media, Istmo crude oil with API index = 33.74 [48], in its heavy fraction (with molecules greater than C18) and the filter paper (squares with 1.5 cm on each side), were sterilized in wet heat during 20 minutes in autoclave at 121°C and 127,486 kPa (1.3 kgf·cm^{-2}) [48]. A 1.5 × 1.5 cm filter paper impregnated with crude oil was placed on the SCA in the petri dishes under axenic conditions. On the other hand, the strains of fungi obtained from the previous phase were sown again. The mycelium of the preserved fungus colonies was touched with a handle and striated in the sterilized PDA medium prepared as indicated by the manufacturer. Then it was incubated at 35°C for 72 h. For sowing in SAC, a slice of fungus with a reproductive structure of the sown strains was extracted with a puncher (0.9 cm in diameter) and was placed on filter paper impregnated with oil, then the Petri dish was covered [40]. Sowing was done separately and in triplicate in the SCA medium for each fungal strain and incubated at 29°C for 6 d, with evaluation every 24 h. The three strains that showed more radial development, H14, H18 and H24 passed to the next phase.

Phase III

Adaptation of fungi in liquid mineral medium Cellulose-Agar (LCA) for fungi to degrade Total Petroleum Hydrocarbons (TPH). Fungi selected in Phase II were sown in the LCA medium. This was prepared as suggested by Rivera-Cruz [40], with crude oil incorporated into the culture medium as carbon and energy source. Then each organism was sown separately, as proposed by Rivera-Cruz et al. [47]. The fungal growth was determined every 72 h for 6 d and H24 strain was selected because of its higher population growth. A micro-culture of H24 in PDA was performed, and with the keys of Barnett and Hunter [49]. The organism was identified as Penicillium sp. This fungus was preserved in PDA medium at 4°C in inclined tube.

Stage 2. Development of Experimental Design to Determine Optimal Growth Parameters

The second stage was conducted in two phases. In the first phase an in vitro bioassay was established where the fungus selected in the previous stage was subjected to four nutrient media, varying pH and temperature. In the second phase a test of fungal biomass production was done in a bioreactor (Kettler jar) with the parameters that stimulated an optimal development of the microorganism.

Phase I

An in vitro bioassay was established, based on a completely randomized design in 3 × 3 × 4 factorial type, namely: three levels of temperature ($T_1 = 29°C$, $T_2 = 35°C$ and $T_3 = 40°C$), three pH levels ($pH_1 = 3.5$, $pH_2 = 5.0$ and $pH_3 = 6.0$) and four types of nutrients (N_1 = Urea, N_2 = Triple-17, N_3 = Nitrophoska-Blue and N_4 = PureSalts), under the statistical model of Equiation (1).

$$Y_{ijk} = \mu + T_i + N_j + pH_k + T_iN_j + T_ipH_k + N_ipH_j + E_{\ell(ijk)} \quad (1)$$

Temperatures of 35°C and 40°C were established using two glass trays, each connected to an electric heater. Water temperature was kept constant using a submersible recirculation pump. To determine nutrients and pH parameters, 250 mL Erlenmeyer flasks were used, divided into three series of 12 treatments; they were connected to a compressor with filter membrane and valve controller for supplying sterile air.

Preparation of culture media. Growth media used were: N_1 = Urea (Abbot), N_2 = Triple-17 NPK (Rancho Los Molinos), N_3 = Nitrophoska-Blue 12-12-17-6 NPKS (Compo) and N_4 = pure-Salts Na_2HPO_4, KH_2PO_4, NH_4Cl, $MgSO_4 \cdot 7H_2O$ [50], prepared with equal proportions of C, based on the same amount of glucose (Merck), added to each medium. The fertilizer nutrients had proportional relations in the following NPK elements: proportion between Pure-Salts: Triple-17, N (approx. 1:1), P (approx. 4:1), K (approx. 2:1), respectively; proportion between Triple-17:Nitrophoska-Blue, P (2.8:1), N (2.8:1), K (1:1); proportion between Urea:Pure-Salts, N (3.4:1). Moreover, the mass load was the same for the three organic fertilizers. Media were prepared as follows: for each of the four media, three Erlenmeyer flasks with 500 mL of distilled water (Mercury Chemical) and 2.5 g of glucose as carbon source were prepared. The first medium contained 0.25 g of Urea, the second: 0.25 g of Triple-17, the third, 0.25 g Nitrophoska-Blue and fourth: 0.5325 g of Na_2HPO_4 (Fermont), 0.325 g of KH_2PO_4 (Fermont), 0.125 g of NH_4Cl (Golden Bell), 0.05 g of $MgSO_4 \cdot 7H_2O$ (JT Baker), making a total of 12 flasks. Then, each of these nutrient media was adjusted in pH to 3.5, 5.0 and 6.0 respectively, using 0.1 M of H_2SO_4 (JT Baker) and NaOH (JT Baker) solutions. To assess the nutrient media at each temperature, 150 mL of each nutrient medium with its pH adjusted were measured in triplicate and placed in a 250 mL flask, obtaining 36 experimental units (EU), according to experimental design. The EU were sterilized for 20 min at 121°C and 103,421 kPa (15 $lb_f \cdot plg^{-2}$). These EU were subsequently inoculated with Penicillium sp.

Preparation of inoculum: The Penicillium sp. Preserved (Phase III in Stage I) was re-isolated in Petri dishes on PDA medium prepared as indicated by the manufacturer, then it was sterilized; the mycelium of the fungus colonies preserved was touched with a handle and striated in the PDA medium. Then, it was incubated at 35°C for 72 h. After that, the mycelium of the re-insulated fungus was touched with a handle and striated in the PDA medium in an inclined tube and incubated at 35°C for 4 d. Then 5 mL of sterile distilled water was added in the inclined tube to allow separation of the spores. The above mixture was poured into an Erlenmeyer flask with 100 mL of water in axenic conditions. The initial spore concentration was determined in this volume by counting in a Neubauer chamber according to Pica et al. [51]. 2.5 mL of this inoculant medium was injected in each of the 36 EU under axenic conditions to homogenize and the initial amount of inoculum Colony Forming Unit (CFU) of Penicillium sp. was determined at time zero hours (t = 0 h). The evaluation of the Penicillium sp. growth in the EU was made every 2 d for 37 d by the method of counting viable cells by plating on surface [46]. Also, pH values and temperature were measured every 24 h during the time the experiment lasted.

Phase II

Production of fungal biomass in a bioreactor (Kettler jar). The production of biomass Penicillium sp. was done in a 2000 mL bioreactor fitted with a mechanical stirrer, sterile air inlet, sample taking, and air outlet to release pressure. The optimal parameters for microbial growth from the previous phase: N_2 = Triple-17, pH_2 =5.0, T_1 = 29°C were adjusted to prepare 1100 mL of nutrient medium that was sterilized in the bioreactor for 20 min at 121°C and 103,421 kPa (15 lb/plg²), and 18.33 mL of inoculant prepared similarly to the previous phase was added. Then, it was shaken to homogenize and sampled to determine initial CFU number (t = 0 h). The evaluation of the development of Penicillium sp. was performed every 24 h for a period of 15 d. In both cases the quantification of CFU of the microorganism is determined similarly to the previous phase. Temperature and pH were measured during each evaluation of fungal growth.

RESULTS

The study was conducted on soil contaminated with Istmo crude oil (hydrocarbon chain greater than C18), with API = 33.74 [48], in order to find and isolate hydrocarbonoclastic microorganisms to be used in the production of fungal biomass. It was determined that the soil was contaminated with 4.0×10^5 mg·kg^{-1} HTP (400,000 ppm). In the microbial isolation (Phase 1 of Stage 1), 24 strains of total fungi were found. Figure 1 shows the results for 14 strains of fungus. The 5 strains selected for further evaluation in Phase II were those with the highest radial development in the PDA culture medium.

Sowing the 5 strains of the previous Phase in Solid medium Cellulose-Agar (SCA) for TPH degrading fungi (Phase II) resulted in a greater development for strains H14, H18 and H24. These 3 strains passed to phase III (Table 1).

Table 2 (Phase III) shows the growth performance of the three selected fungi in Phase II in a mineral medium enriched with crude oil (LCA). Its speed behavior was observed over time (6 d). Likewise, it was verified that the crude oil surface tension cracked, completely changing its appearance. With these results, the three strains of fungi were identified as hydrocarbonoclastic. From these strains, the one named H24 and identified as Penicillium sp. by comparison with the keys of Barnett and Hunter [49] had the largest growth during the period of the experiment and was selected for evaluation in Stage 2.

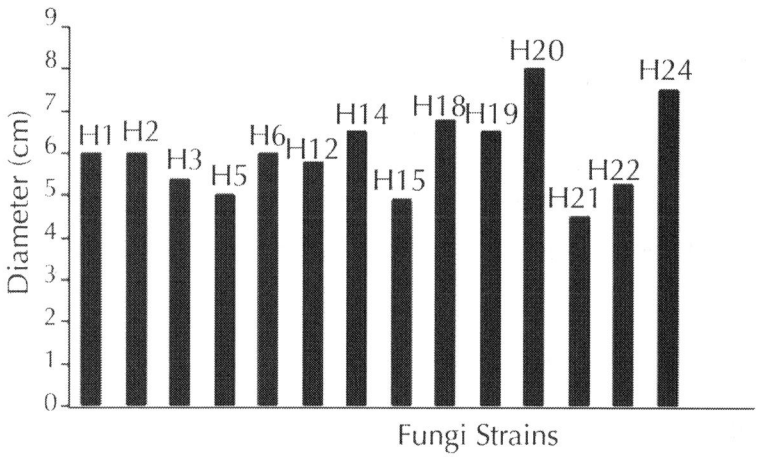

Figure 1: Radial growth of the fungi on PDA medium.

Table 1: Radial growth of fungal strains in Solid medium Cellulose-Agar (SCA) to TPH degrading fungi

Fungal Strain	Development radial (cm) on each day (d)					
	1 d	2 d	3 d	4 d	5 d	6 d
H14	0.9	1.1	2.0	3.2	4.6	5.4
H18	0.9	1.3	2.2	3.5	4.7	6.2
H19	0.9	0.9	1.2	1.5	1.8	1.9
H20	0.9	1.0	1.5	1.8	2.0	2.2
H24	0.9	1.4	2.5	3.8	5.5	7.0

Table 2: Growth of fungal strains as colony forming units (CFU) in Liquid mineral medium Cellulose-Agar (LCA), enriched with crude oil during the 6-d-long experiment

Fungal Strain	CFU start t = 0 h	CFU at t = 72 h	CFU at a t = 144 h
Control (without oil)	3.0×10^{-3}	3.1×10^{-3}	2.8×10^{-3}
H14	3.0×10^{-3}	2.8×10^{-4}	3.1×10^{-5}
H18	3.0×10^{-3}	3.4×10^{-4}	5.2×10^{-5}
H24	3.0×10^{-3}	7.2×10^{-4}	7.8×10^{-4}

In assessing the growth of Penicillium sp. in a flask (Stage 2, Phase I), an increase in the viscosity of the medium during the biomass growth was observed. With the data obtained the mean test was done.

Figure 2 shows the results of treatments at different temperatures, nutrients and pH. Figure 2(a) indicates that the growth of Penicillium sp. was the best at a temperature of 29°C, but slightly less in the other two temperatures. Figure 2(b) indicates that the microorganism had a better development in the medium with Pure-Salts, but not very different from the media with NitrophoskaBlue and Triple-17. The lowest growth was in the medium with Urea. Figure 2(c) shows that the population growth of Penicillium sp. was higher at pH 3.5 but not so different to that achieved in the other two pH values.

Figure 3 shows the results of the treatments that had the largest population growth of Penicillium sp. According to the experimental design. The amount of microorganisms to t = 0 h was 4500 CFU·mL^{-1}. It was observed that the best results were obtained with the medium N_4 = Pure-Salts, pH_1 = 3.5, T_1 = 29°C, with a maximum population growth of 2.94×10^6 CFU·mL^{-1} at 13 d. The next best treatments were (in decreasing order of fungal growth): N_2 = Triple-17, pH_2 = 5.0, T_1 = 29°C (2.06×10^6 CFU·mL^{-1} at 11 d), N_2 = Triple-17, pH_1 = 3.5, T_1 = 29°C (1.50×10^5 CFU·mL^{-1} at 25 d) and N_2 = Triple-17 pH_3 = 6.0, T_2 = 35°C (2.85×10^6 at 20 d), respectively. The maximum multiplication of the media prepared with urea was N_1 = Urea, pH_1 = 3.5, T_1 = 29°C (8.9×10^4 CFU·mL^{-1}

at 25 d), being significantly lower compared to the previous ones.

The design for the production of fungal biomass from Penicillium sp. in the bioreactor (Phase II of Stage II) considered both, the best experimental treatment ($N_4pH_1T_1$), and the most suitable organic media ($N_2pH_2T_1$). The assessment of the population dynamics of the fungal strain in the bioreactor was performed in the selected medium N_2 = Triple-17, pH_2 = 5.0, T_1 = 29°C, because it proved to be the most favorable among all organic nutrient media and economically, the most suitable of all evaluated media for possible application to larger scales. Thus, in the medium $N_2pH_2T_1$ the exponential phase of fungal growth was observed between days 8 to 10 of culture, peaking on day 10 to 2.1×10^6 CFU·mL^{-1}, to minimally decrease later the cell number (Figure 4).

DISCUSSION

The study was conducted in the state of Tabasco, Mexico on a Gleysol-mollic soil that has been contaminated for more than 20 years with crude oil [52]. A value of 4.0×10^5 mg·kg^{-1} TPH was found (400,000 ppm). 24 strains of total fungi were obtained (Figure 1 shows the 14 strains with higher multiplication). The so-called H14, H18, H19, H29 and H24 had the highest population development in the PDA medium and then were sown in the SCA medium where colonies H14, H18 and H24 showed greater growth (Table 1). They could be considered, according to Atlas et al. [53] and Adams-Schroeder et al. [54], as specialized degrading hydrocarbons strains, because of their ability to adapt in soils with high concentrations of crude oil and use it as a source of carbon and energy due to their genetic potential [16,25,46,55,56]. Subsequently, these 3 strains were sown in the LCA medium, where H24 had the highest population growth (Table 2), and was classified as Penicillium sp. based on macro and microscopic characteristics [49,57-59].

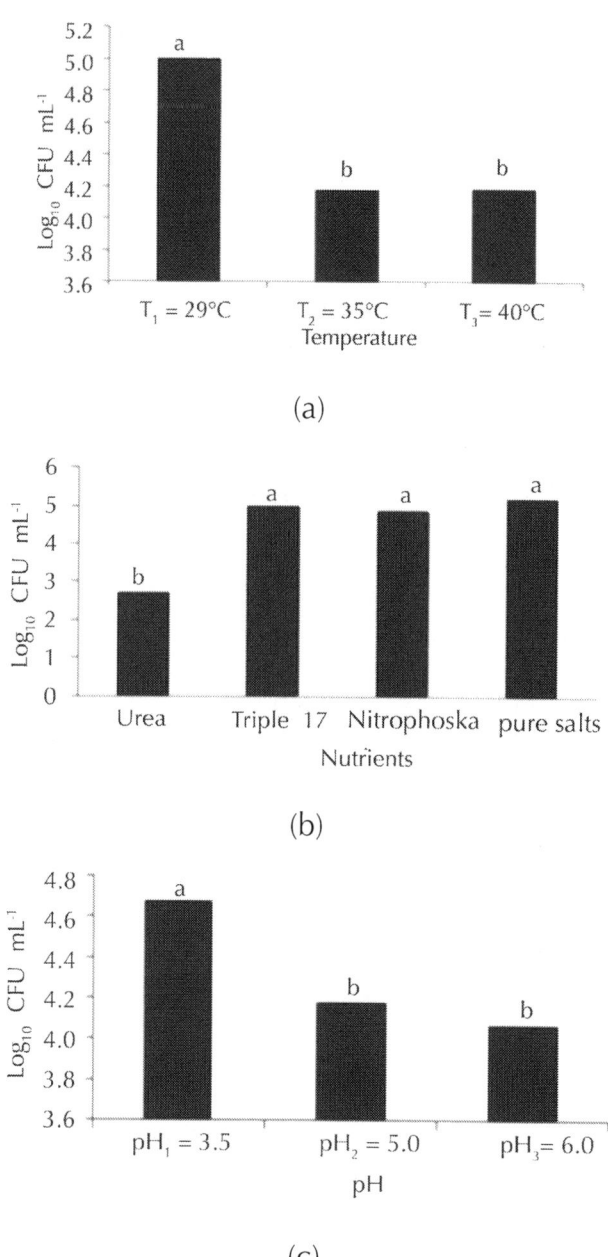

Figure 2: Growth of Penicillium sp. at different conditions of: (a) Temperature, (b) Nutrients and (c) pH. Treatment means with different letters are statistically significant differences (≤ 0.05).

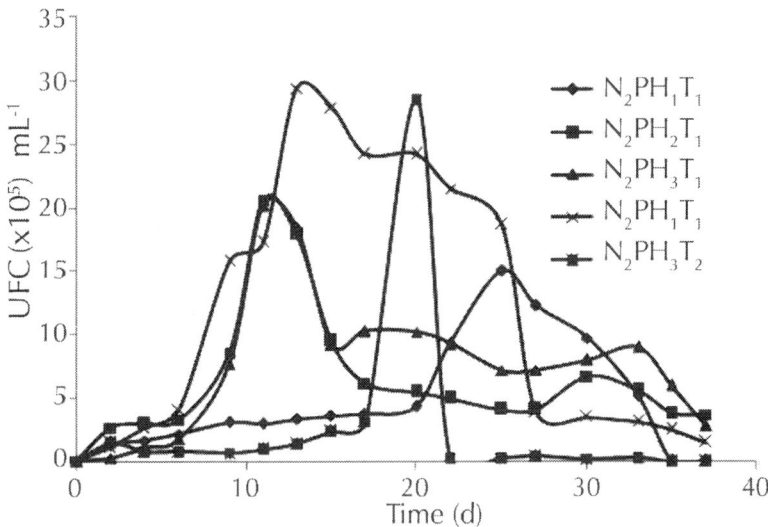

Figure 3: Growth of Penicillium sp. on flask treatments (4.5 × 10³CFU·mL⁻¹, to t = 0 d).

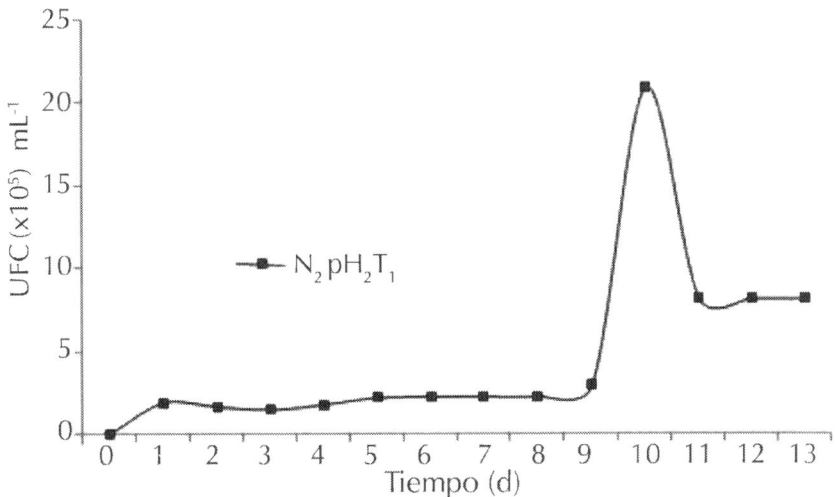

Figure 4: Growth of Penicillium sp. on bioreactor treatment (4.2 × 10³CFU·mL⁻¹, to t = 0 d).

These findings were consistent with those found in other countries with oil industry, since the Penicillium fungi have been found in soils in kwait [60], Canada [61], Africa [16,17,29,62], Chile [39] and Korea [63], among others. Even in studies of agricultural soils in their natural state, it was found that native Penicillium sp. survived and degraded hydrocarbons when it was added crude oil, without having previously been in contact with the pollutant [28]. Likewise, when Penicillium sp. from feces of cattle and poultry was added on contaminated soil it survived and degraded crude oil, although just participating in fungal consortia [64]. Several studies have indicated that the Penicillium sp. degraded crude oil, using it as its sole source of carbon and energy [16,23,25,26,28,64-66], or oil derivates with the same purpose [9,25,29,31,63]. This can be attributed to the fact that Penicillium sp. has in its membrane a specific group of oxygenases associated with the production of surface-active agents [67,68], which reduce the surface tension of the aqueous medium increasing the TPH solubility [69]. In this regard, we can only discuss so far the potential of Penicillium in the field of bioremediation, since large-scale studies have not been reported [25]. Nor information is available on fungal biomass generation of this hydrocarbonoclastic fungus and its application in bioremediation of crude oil in soils.

Afterward, the experiment was carried out in laboratory to determine the optimal parameters of fungal growth. The Penicillium sp. showed red pigmentation in the treatments with Urea. Mendez-Zavala et al. [70], noted a similar color with Penicillium purpurogenum. This might be because the Urea treatments contained a greater concentration of nitrogen. Cho et al. [71] reported that the nitrogen source marks a very significant effect on the expression of pigments as well as the presence of sucrose and starch [72]. Alexander [73] indicated that the organisms have optimal growth in a temperature range of 28°C to 30°C, although Saraswathy et al. [74], used 22°C in their studies with Penicillium ochrochloron. Several researchers have reported 30°C or 31°C as optimal temperatures for development of Penicillium sp. [9,23,25,26]. Corry [75] and Lacey [76] have also reported that nine species of Penicillium have an average temperature of 28°C for optimum growth. Figure 2(a) shows that Penicillium sp. had the best population growth at a temperature of 29°C (Tukey ≤ 0.05), which agrees with Isitua and Ibeh [29], who have recommended an interval of 28°C ± 2°C for the optimum growth of Penicillium sp. With regard to nutrients, significant statistical evidence (Figure 2(b)) supports the fact that the best growth was given in the Pure-Salts treatment (1.6×10^5 CFU·mL^{-1}); a reason could be that it contains N, P and K and other trace elements. According to what Ercoli et al. report [77], the source of trace elements helps to improve the process of biodegradation of hydrocarbons. NitrophoskaBlue, Triple-17 and Urea fertilizers do not contain these bioelements. Figure 2(c) shows a statistically significant difference (Tukey ≤ 0.05), with the largest population of 1×10^5 CFU·mL^{-1} in the treatment $pH_1 = 3.5$. According to Alexander [73], fungi have optimum growth in agar medium with a pH of 4.0 approximately. The results obtained by Sanchis et al. [78], in a study of Penicillium griseofulvum indicated that the optimum pH is reached in a range of 3.5 to 4.5. Once with these data, the experimental design was outlined to determine the best conditions for Penicillium sp. growth, and so use the most efficient treatment in obtaining hydrocarbonoclastic fungal biomass.

Figure 3 shows the fungal growth obtained in the 5 best treatments of the experimental design. All media had a rapid microbial growth the first two days, and then continue their multiplication in the same order of magnitude gradually, prior to exponential growth. The treatment that achieved the highest multiplication of Penicillium sp. was the one formulated with Pure Salts $N_4pH_1T_1$, while the best organic media were those prepared with the Triple-17 fertilizer, and among them, the most efficient was $N_2pH_2T_1$ with a gradual growth of the microorganism to 4.2×10^5 CFU·mL^{-1} on day 6. Then it multiplied exponentially to reach 2.06×10^6 CFU·mL^{-1} at 11 d, decreased to 6.2×10^5 CFU·mL^{-1} at day 17 and then it maintained a plateau with little variation in fungal concentration, to reach 3.7×10^5 at the end of the experiment (37 d). On the other hand, the medium $N_2pH_3T_2$ had a long multiplication time until day 17 (3.0×10^5 CFU·mL^{-1}), reaching the maximum exponential growth on day

20 with 2.85×10^6 CFU·mL^{-1} and then it decreased sharply to 3.0×10^4 on day 22 and maintained a plateau with little change until day 33 (3.0×10^4 CFU·mL^{-1}). The medium $N_2pH_1T_1$ had a gradual growth of 1.3×10^5 to 3.8×10^5 CFU·mL^{-1} from day 2 to 17, and then an exponential growth reaching a maximum concentration at day 25 with 1.5×10^6 CFU·mL^{-1}; then it decreased to 1.0×10^4 CFU·mL^{-1} at 37 d without forming a plateau. Moreover, the development of the fungus in the medium $N_1pH_1T_1$ remained in the same order of magnitude, with 1.5 to 2.9×10^4 CFU·mL^{-1} from 2 to 17, then increased to a maximum population of 8.9×10^4 CFU·mL^{-1} to day 25 and then decreased to 1.4×10^4 CFU·mL^{-1} at day 33. $N_2pH_2T_1$ treatment reached a maximum population development 27.7% less than the one obtained by the $N_2pH_3T_2$, but the first was achieved significantly faster (11 d and 20 d, respectively).

Comparing treatments $N_2pH_2T_1$ and $N_2pH_1T_1$, the second had a maximum population growth of 7.3% of the first, while treatment $N_1pH_1T_1$ had a multiplication of only 4.3% compared to $N_2pH_2T_1$; also the times to reach maximum microbial development was considerably lower for the $N_2pH_2T_1$. The experimental conditions of $N_2pH_2T_1$ treatment are closely related to the pH and temperature values found as optimal in the literature as described above. It also provides the shortest adaptation time of Penicillium sp. and reaches a plateau of considerable concentration of microorganisms for a long time after the maximum fungal multiplication. The lowest microorganism reproduction was observed in media with urea. Likewise, the cost in Mexico (50 kg package, in U.S. Dollar) for each fertilizer used in this study was as follows: Urea $21.34, Triple-17 $25.77, NitrophoskaBlue $61.04, and for Pure Salts: Na_2HPO_4 $24.15 (250 g), KH_2PO_4 $28.66 (250 g), NH_4Cl $48.31 (500 g), $MgSO_4 \cdot 7H_2O$ $44.28 (500 g).

Based on the above, we chose the medium $N_2pH_2T_1$ to obtain fungal biomass in the bioreactor, with the additional consideration that Triple-17 fertilizer compared to Nitrophoska-Blue has a higher proportion of N and P and an equal proportion of K (in addition to a better fungal growth). Besides, it is cheaper than Nitrophoska-Blue and the medium prepared with Pure Salts. Triple-17 was considered better than Urea because of its highest multiplication of Penicillium sp. and the high proportion of NPK nutrients that it has, although it was slightly more expensive than Urea.

The evaluation of the multiplication of Penicillium sp. in the bioreactor was carried out only after completing the stage of maximum microbial growth and the beginning of the decrease in concentration of fungi because there was an interest in obtaining the maximum fungal biomass. Comparing the results obtained with the flask and those obtained in the bioreactor about the growth assessment of the hydrocarbonoclastic fungus Penicillium sp. in the selected medium $N_2pH_2T_1$ prepared with Triple-17 (Figures 3 and 4), it was found that in a bioreactor a maximum fungal proliferation of 2.1×10^6 was obtained at day 10, which means an increase of 2% in the $CFU \cdot mL^{-1}$ of the organism with respect to the experiment in the flask, decreasing thereafter to 8.17×10^5 at day 13. It was noted that the growth curve in both cases (flask and bioreactor) was not exactly the same, although very similar in behavior, with reduced time to obtain maximum fungal population from 11 to 10 d in the bioreactor. That could be due to greater efficiency in managing and controlling the aeration of the medium during the duration of the experiment in contrast to the study in the flask. The concentration of microorganisms reached in the bioreactor was high enough to be considered for possible inoculation of soil in bioremediation projects, compared to data published by Rivera-Cruz [40], who found low values like 10^3 CFU of fungi in the rhizosphere of German grass (Echinochloa polistachya) in Gleysol soil contaminated with crude oil.

CONCLUSIONS

It was determined that a Gleysol-mollic soil, in the state of Tabasco, México, with over 20 years of being impacted by crude oil, was contaminated with 4.0×10^5 $mg \cdot kg^{-1}$ (400,000 ppm) of TPH. From this soil, the hydrocarbonoclastic fungus Penicillium sp. was isolated and characterized. This organism was adapted to various conditions of temperature, pH and four nutrient media. Considering the variables listed in the experimental design and costs of the materials used to prepare the nutrient media, it was found that the treatment at a temperature of 29°C, pH 5.0, and nutrient media Triple-17 ($N_2pH_2T_1$) had the highest population growth at 11 d with $2.06 \times 10^6 CFU \cdot mL^{-1}$. Subsequently, the experiment was performed in a bioreactor for the production of fungal hydrocarbonoclastic biomass, and it was found that the proliferation of Penicillium sp. was obtained at 10 d with 2.1×10^6 CFU·mL^{-1}. If this development could be achieved in contaminated soils with the proper soil preparation and bioaugmentation techniques, it would be considered a potential bioremediation strategy for solving the problem of hydrocarbon-contaminated soils in Tabasco in particular and in general in Mexico.

ACKNOWLEDGMENTS

This research is part of the POA-2008011 project, "Determining of the optimal parameters to produce Fungal and Bacterial hydrocarbonoclastic Biomass" developed by the División Académica de Ingeniería y Arquitectura (DAIA) of the Universidad Juárez Autónoma de Tabasco (UJAT). It receives partial funding by the company Corporativo de Servicios Ambientales S. A. de C. V. (CORSA). The authors thank the DAIA/UJAT for all the support to carry out this research, to Ing. Alfredo Castro Betancourt, General Manager of CORSA S. A. de C. V. for all the procedures he went through to obtain finance for this project, to Mr. Alex Figueroa Munóz and Eva Flandes Aguilera, Dean of the Facultad de Salud y Ciencias A. F. and academic vice-rector, respectively, of the Universidad SEK, for their support in the organization and conclusion of this document.

REFERENCES

1. INEGI, "Comunicado Núm. 203/09," INEGI, 2009.

2. N. Cruz-Serrano, "Pemex, tercera petrolera en el mundo en 2009," El Universal, 2013. http://www.eluniversal.com.mx/finanzas/80420.html
3. E. C. Hernández, "Innovación Tecnológica Base Para Extracción de Petróleo," Milenio Tabasco, 2013. http://www.skyscrapercity.com/showthread.php?t=634607&page=18
4. S. Arias, "Desarrollo Económico de Tabasco," Tabasco hoy, 2013. http://www.tabascohoy.com.mx/noticia.php?id_nota=190648
5. M. Levin and M. Gealt, "Biotratamiento de Residuos Tóxicos y Peligrosos," McGraw-Hill. Madrid, 1997.
6. J. Eweis, S. Ergas, D. Chag and E. Schoroeder, "Principios de Biorrecuperación," McGraw-Hill, Madrid, 1999.
7. K. Watanabe, "Microorganisms Relevant to Bioremediation," Current Opinion in Biotechnology, Vol. 12, No. 3, 2001, pp. 237-241. http://dx.doi.org/10.1016/S0958-1669(00)00205-6
8. W. T. Jr. Frankenberger, "The Need for a Laboratory Feasibility Study in Bioremediation of Petroleum Hydrocarbons," In: E. J. Calabrese andP. T. Kostecki, Eds., Hydrocarbon Contaminated Soils and Groundwater, Lewis Publication, Boca Raton, 1992, pp. 237-293.
9. Y. Castro-Riquelme, "Estudios de Toxicidad y Biodegradacion de Hidrocarburos Modelo en Hongos Filamentosos," Maestría en Biotecnología Dissertation. Universidad Autónoma Metropolitana, México, D.F., 2008.
10. M. V. Walter, "Bioaugmentation," In: C. J. Hurst, Ed., Manual of Environmental Microbiology, ASM Press, Washington DC, 1997, pp. 753-765.
11. R. M. Atlas and R. Unterman, "Bioremediation," In: A. L. Demain and J. E. Davies, Eds., Manual of Industrial Microbiology and Biotechnology, 2nd Edition, ASM Press, Washington DC, 1999, pp. 666-681.
12. S. Boonchan, M. L. Britz and G. A. Stanley, "Degradation and Mineralization of High-Molecular-Weight Polycyclic Aromatic Hydrocarbons by Defined FungalBacterial

Cocultures," Applied and Environmental Microbiology, Vol. 66, No. 3, 2000, pp. 1007-1019. http://dx.doi.org/10.1128/AEM.66.3.1007-1019.2000

13. S. Barathi and N. Vasudevan, "Utilization of Petroleum Hydrocarbons by Pseudomonas fluorescens Isolated from a Petroleum-Contaminated Soil," Environment International, Vol. 26, No. 5-6, 2001, pp. 413-416. http://dx.doi.org/10.1016/S0160-4120(01)00021-6

14. E. Seklemova, A. Pavlova and K. Kovacheva, "Biostimulation Based Bioremediation of Diesel Fuel: Field Demonstration," Biodegradation, Vol. 12, No. 5, 2001, pp. 311-316. http://dx.doi.org/10.1023/A:1014356223118

15. I. Kuiper, E. L. Lagendijk, G. O. Bloemberg and B. J. J. Lugtenberg, "Rhizoremediation. A Beneficial Plant Microbe Interaction," Molecular Plant-Microbe Interactions, Vol. 17, No. 1, 2004, pp. 6-15. http://dx.doi.org/10.1094/MPMI.2004.17.1.6

16. T. E. Ogbulie, H. C. Nwigwe, M. O. E. Iwuala and G. C. Okpokwasili, "Study on the Use of Monoculture and Multispecies on Bioaugumentation of Crude Oil Contaminated Agricultural Soil," Nigerian Journal of Microbiology, Vol. 24, 2010, pp. 2160-2167.

17. A. F. Gesinde, E. B. Agbo, M. O. Agho and E. F. C. Dike, "Bioremediation of Some Nigerian and Arabian Crude Oils by Fungal Isolates," International Journal of Pure and Applied Sciences, Vol. 2, No. 3, 2008, pp. 37-44.

18. A. D'Annibale, F. Rosetto, V. Leonardi, F. Federici and M. Petruccioli, "Role of Autochthonous Filamentous Fungi in Bioremediation of a Soil Historically Contaminated with Aromatic Hydrocarbons," Applied and Environmental Microbiology, Vol. 72, No. 1, 2006, pp. 28-36. http://dx.doi.org/10.1128/AEM.72.1.28-36.2006

19. Conestoga-Rovers & Associates (CRA), "Bioaugmentation," Innovative Technology Group, Vol. 3, No. 4, 2003, pp. 1-2.

20. J. B. Davis, "Petroleum Microbiology," Elsevier, Amsterdam, 1967.
21. J. A. Bumpus, "Biodegradation of Polycyclic Aromatic Hydrocarbons by Phanerochaete chrysosporium," Applied and Environmental Microbiology, Vol. 55, No. 1, 1989, pp. 154-158.
22. T. S. Brodkorb and R. L. Legge, "Enhanced Biodegradation of Phenanthrene in Oil Tar-Contaminated Soils Supplemented with Phanerochaete chrysosporium," Applied and Environmental Microbiology, Vol. 58, No. 9, 1992, pp. 3117-3121.
23. J. L. Solórzano-Lemos, A. C. Rizzo, V. S. Millioli, A. U. Soriano, M. I. De Moura-Sarquis and R. Santos, "Petroleum Degradation by Filamentous Fungi," Contribuição Técnica a 9th International Petroleum Environmental Conference, Novo México, 2002, pp. 21-25.
24. H. Sing, "Mycoremediation," John Wiley & Sons, Inc., Hoboken, 2006. http://dx.doi.org/10.1002/0470050594
25. A. L. Leitão, "Potential of Penicillium Species in the Bioremediation Field," International Journal of Environmental Research and Public Health, Vol. 6, No. 4, 2009, pp. 1393-1417. http://dx.doi.org/10.3390/ijerph6041393
26. H. M. Hussein and Y. R. Abdel-Fattah, "Numerical Modelling of Petroleum Oil Bioremediation by a Local Penicillium Isolate as Affected with Culture Conditions: Application of Plackett-Burman Design," Arab Journal of Biotechnology, Vol. 5, No. 2, 2002, pp. 165-172.
27. A. Mittal and P. Singh, "Studies on Biodegradation of Crude Oil by Aspergillus niger," The South Pacific Journal of Natural and Applied Sciences, Vol. 27, No. 1, 2009, pp. 57-60.
28. O. Obire and E. C. Anyanwu, "Impact of Various Concentrations of Crude Oil on Fungal Populations of Soil," International Journal of Environmental Science and Technology, Vol. 6, No. 2, 2009, pp. 211-218.

29. C. C. Isitua and I. N. Ibeh, "Comparative Study of Aspergillus niger and Penicillium sp. in the Biodegradation of Automotive Gas Oil (AGO) and Premium Motor Spirit (PMS)," African Journal of Biotechnology, Vol. 9, 2010, pp. 3607-3610.
30. E. Kiehlmann, L. Pinto and M. Moore, "The Transformation of Chrysene to Trans-1,2-dihydroxy-1,2-dihydrochrysene by Filamentous Fungi," Canadian Journal of Microbiology, Vol. 42, 1996, pp. 604-608. http://dx.doi.org/10.1139/m96-081
31. C. Machín-Ramírez, D. Morales, F. Martínez-Morales, A. I. Okoh and M. R. Trejo-Hernández, "Benzo[a]pyrene removal by Axenicand Co-Cultures of Some Bacterial and Fungal Strains," International Biodeterioration and Biodegradation, Vol. 64, No. 7, 2010, pp. 538-544. http://dx.doi.org/10.1016/j.ibiod.2010.05.006
32. H. J. Van den Brink, R. F. M. van Gorcom, C. A. M. J. J. van den Hondel and P. J. Punt, "Cytochrome P450 Enzyme Systems in Fungi," Fungal Genetics and Biology, Vol. 23, No. 1, 1998, pp. 1-17. http://dx.doi.org/10.1006/fgbi.1997.1021
33. C. E. Cerniglia and J. B. Sutherland, "Bioremediation of Polycyclic Aromatic Hydrocarbons by Ligninolytic Fungi," In: G. M. Gadd, Ed., Fungi in Bioremediation, Cambridge University Press, Cambridge, 2001, pp. 136-187. http://dx.doi.org/10.1017/CBO9780511541780.008
34. J. A. Field, E. Jong, G. F. Cost and J. A. M. Bont, "Biodegradation of Polycyclic Aromatic Hydrocarbons by New Isolates of White Rot Fungi," Applied and Environmental Microbiology, Vol. 58, No. 7, 1992, pp. 2219-2226.
35. L. Launen, L. Pinto, C. Wiebe, E. Kiehlmann and M. Moore, "The Oxidation of Pyrene and Benzo[a]pyrene by nonbasidiomycete Soil Fungi," Canadian Journal of Microbiology, Vol. 41, No. 6, 1995, pp. 477-488. http://dx.doi.org/10.1139/m95-064
36. L. Launen, L. Pinto and M. Moore, "Optimization of Pyrene Oxidation by Penicillium janthinellum Using Response-Surface Methodology," Applied Microbiology and

Biotechnology, Vol. 51, No. 4, 1999, pp. 510-515. http://dx.doi.org/10.1007/s002530051425

37. T. Wunder, J. Marr, S. Kremer, O. Sterner and H. Anke, "1-Methoxypyrene and 1,6-Dimethoxypyrene: Two Novel Metabolites in Fungal Metabolism of Polycyclic Aromatic Hydrocarbons," Archives of Microbiology, Vol. 167, 1997, No. 5, pp. 310-316. http://dx.doi.org/10.1007/s002030050449

38. B. Chávez-Gómez, R. Quintero, F. Esparza-García, A. M. Mesta-Howard, D. F. J. Zavala, C. H. HernándezRodríguez, T. Gillén, H. M. Poggi-Varaldo, J. BarreraCortés and R. Rodríguez-Vázquez, "Removal of Phenanthrene from Soil by Co-Cultures of Bacteria and Fungi Pregrown on Sugarcane Bagasse Pith," Bioresource Technology, Vol. 89, No. 2, 2003, pp. 177-183. http://dx.doi.org/10.1016/S0960-8524(03)00037-3

39. F. E. Valenzuela, M. L. Solís, V. O. Martínez and T. D. Pinochet, "Hongos Aislados Desde Suelos Contaminados Con Petróleo," Boletín Micológico, Vol. 21, 2006, pp. 35-41.

40. M. C. Rivera-Cruz, "Microorganismos Rizosféricos de Los Pastos Alemán (Echinochloa polystachya H.B.K. Hitchc) y Cabezón (Paspahum virgatum L.) en la Degradación Del Petróleo Crudo y el Benzo(a)pireno," Ph.D. Dissertation, Colegio de Postgraduados, Montecillo, 2001.

41. INEGI, "Síntesis Geográfica, Nomenclátor y Anexo Cartográfico del Estado de Tabasco," Instituto Nacional de Estadística y Geografía, Aguascalientes, 2001.

42. SEMARNAT, "Norma Oficial Mexicana NOM-021-RECNAT-2000. Apartado 6.1. Evaluación de la Conformidad Para Muestreo de Suelos. Muestreo Para Determinar Fertilidad de Suelos," Diario Oficial de la Federación, 2nd. Secc., México D.F., 2002.

43. USEPA, "EPA-Method-3540C. Soxhlet Extraction. Hidrocarburos Totales Del Petróleo (Fracción Pesada)," 1996. http://www.epa.gov/wastes/hazard/testmethods/sw846/pdfs/3540c.pdf

44. USEPA, "EPA-Method-9071B. n-Hexane Extractable Material (hem) for Sludge, Sediment, and Solid Samples," 1998. http://www.caslab.com/EPA-Methods/PDF/EPA-Method-9071B.pdf
45. USEPA, "EPA-Method-1664A. Revition A. n-Hexane Extractable Material," 1999.http://www.epa.gov/waterscience/methods/method/oil/1664guide.pdf
46. M. T. Madigan, J. M. Martinko, P. V. Dunlap and D. P. Clark, "Biología de Los Microorganismos," Pearson, Addison Wesley, Madrid, 2009.
47. M. C. Rivera-Cruz, R. Ferrera-Cerrato, R. RodríguezVásquez and L. Fernández-Linares, "Adaptación y Selección de Microorganismos Autóctonos en Médios de Cultivos Enriquecidos Com Petróleo Crudo," Terra Latinoamericana, Vol. 20, No. 4, 2002, pp. 423-444.
48. H. Corvantes, "PEMEX: El Petróleo," Petróleos Mexicanos, 1988.
49. H. Barnett and B. Hunter, "Illustrated Genera of Imperfect Fungi," Burgess Pub. Company, Minnesota, 1972.
50. M. Kästner, M. Breuer-Jammali and B. Mhro, "Enumeration and Characterization of the Soil Sites to Mineralize Polycyclie Aromatic Hidrocarbons," Applied Microbiology and Biotechnology, Vol. 41, No. 2, 1994, pp. 267- 273.http://dx.doi.org/10.1007/BF00186971
51. G. Y. Pica, A. Ronco and B. M. Díaz, "Bioensayo de Toxicidad Crónica con Selenastrum capricornutum (Pseudokirchneriella subcapitata). Método de Enumeración Celular Basado en el Uso de Hematocímetro Neubauer," In: M. G. Castillo, Ed., Ensayos Toxicológicos y Métodos de Evaluación de Calidad de Aguas, Estandarización, intercalibración, Resultados y Aplicaciones, Instituto Mexicano de Tecnología del Agua, 2004, pp. 62-73.
52. J. Zavala-Cruz, V. A. Botello, S. R. H. Adams and A. Ruiz-Bello, "Hidrocarburos Alifáticos y Aromáticos en las Tierras," In: J. Zavala-Cruz, M. C. GutiérrezCastorena and D. J. Palma-López, Eds., Impacto Ambiental en Las Tierras Del Campo

Petrolero Samaria, Colegio de Postgraduados, CONACYT, CCYTET, Villahermosa, Tabasco, 2003, pp. 131-140.
53. M. R. Atlas, A. Horowitz, M. Krichevky and K. A. Bej, "Response of Microbial Population to Environmental Disturbance," Microbial Ecology, Vol. 22, No. 1, 1991, pp. 249-256. http://dx.doi.org/10.1007/BF02540227
54. R. H. Adams-Schroeder, V. I. Domínguez-Rodríguez and L. Vinalay-Carrillo, "Evaluation of Microbial Respiration and Ecotoxicity in Contaminated Soils Representative of the Petroleum Producing Region of Southeastern México," Terra, Vol. 20, No. 3, 2002, pp. 253-265.
55. C. E. Cerniglia and M. A. Heitkamp, "Microbial Degradation of Polycyclic Aromatic Hydrocarbon (PAH) in the Aquatic Environment," In: U. Varausi, Ed., Metabolism of PAH in the Aquatic Environment, CRC Press Inc., Boca Raton, 1987, pp. 41-68.
56. J. E. Heidelberg, I. I. Paulsen, K. E. Nelson, E. J. Gaidos, W. C. Nelson, T. D. Read and J. A. Eison, "Genome Sequence of the Dissimilatory Metal Ion-Reducing Bacterium. Shewanella oneidensis," Nature and Biotechnology, Vol. 1, 2002, pp. 1-6.
57. J. Gilman, "Manual de Los Hongos Del Suelo," Compañía Editorial Continental S.A, México D.F., 1963.
58. G. Smith, "Introducción a la Micología Industrial," Editorial Acribia, España, 1963.
59. T. Mier, C. Toriello and M. Ulloa, "Hongos Microscópicos Saprobios y Parásitos: Métodos de Laboratorio," UAM-UNAM, México D. F., 2000.
60. S. S. Radwan, N. A. Sorkhoh, F. Fardoun and R. H. Al-Hasan, "Soil Management Enhancing Hydrocarbon Degradation in the Polluted Kuwaitii Desert," Applied Microbiology and Biotechnology, Vol. 44, No. 1-2, 1995, pp. 265-270. http://dx.doi.org/10.1007/BF00164513
61. T. M. April, J. M. Foght and R. S. Currah, "Hydrocarbon-Dagrading Filamentous Fungi. Isolated from Flare Pit Soils

in Northern and Western Canada," Canadian Journal of Microbiology, Vol. 46, No. 1, 2000, pp. 38-49.
62. G. Nkwelang, F. L. Kamga, E. Nkeng and S. P. Antai, "Studies on the Diversity, Abundance and Succession of Hydrocarbon Utilizing Micro Organisms in Tropical Soil Polluted with Oily Sludge," African Journal of Biotechnology, Vol. 8, 2008, pp. 1075-1080.
63. K. Jeonge-Dong and L. Choul-Gyun, "Microbial Degradation of Polycyclic Aromatic Hydrocarbons in Soil by Bacterium-Fungus Co-Cultures," Biotechnology and Bioprocess Engineering, Vol. 12, No. 4, 2007, pp. 410-416. http://dx.doi.org/10.1007/BF02931064
64. O. Obire, E. C. Anyanwu and R. N. Okigbo, "Saprophytic and Crude Oil Degrading Fungi from Cow Dung and Poultry Droppings as Bioremediating Agents," Journal of Agricultural Science and Technology, Vol. 4, 2008, pp. 81-89.
65. M. A. Hernández-Rivera, M. E. Ojeda-Morales, J. G. Martínez-Vázquez, V. Villegas-Cornelio and Y. CórdovaBautista, "Optimal Parameters for in Vitro Development of the Hydrocarbonoclastic Microorganism Proteus sp.," Journal of Soil Science and Plant Nutrition, Vol. 11, No. 1, 2011, pp. 29-43.
66. C. C. Okoro, "Biodegradation of Hydrocarbons in Untreated Produce Water Using Pure Fungal Cultures," African Journal of Microbiology Research, Vol. 2, 2008, pp. 217-223.
67. K. Lee and E. M. Levy, "Biodegradation of Petroleum in the Marine Environment and Its Enhancement," In: J. O. Nriagu and J. S. S. Lakshminarayana, Eds., Aquatic Toxicology and Water Quality Management, John Wiley and Sons Inc., New York, 1989, pp. 217-243.
68. B. T. Walton, E. A. Guthrie and A. M. Hoylman, "Toxicant Degradation in the Rhizosphere," In: T. A. Anderson, and J. R. Coats, Eds., Bioremediation through Rhizosphere Technology, American Chemical Society, Washington DC, 1994, pp. 11-26. http://dx.doi.org/10.1021/bk-1994-0563.ch002

69. E. Rosenberg and E. Z. Ron, "Bioremediation of Petroleum Contamination," In: R. L. Crawford and D. L. Crawford, Eds., Bioremediation: Principles and Applications, Cambridge University Press, Cambridge, 1998, pp. 100-124.
70. J. C. Méndez-Zavala, F. Contreras, R. Lara, R. Victoriano and R. Rodríguez, "Producción Fúngica de un Pigmento Rojo Empleando la Cepa Xerofilica Penicillium purpurogenum GH-2," Revista Mexicana de Ingeniería Quí- mica, Vol. 6, 2007, pp. 267-273.
71. Y. J. Cho, J. P. Park, H. J. Hwan, S. W. Kim, J. W. Choi and J. W. Yun, "Production of Red Pigment Bye Submerged Culture of Paecilomyces sinclairii," Letters in Applied Microbiology, Vol. 35, No. 3, 2002, pp. 195-202. http://dx.doi.org/10.1046/j.1472-765X.2002.01168.x
72. Y. J. Cho, H. J. Hwang, S. N. Kim, C. H. Song and J. W. Yun, "Effect of Carbon Source and Aeration Rate on Broth Rheology and Fungal Morphology during Red Pigment Production by Sinclairii in Batch Bioreactor," Journal of Biotechnology, Vol. 95, No. 1, 2002, pp. 13- 23.
73. M. Alexander, "Biodegradation and Bioremediation," Academic Press, San Diego, 1994.
74. A. Saraswathy and R. Hallberg, "Mycelial Pellet Formation by Penicillium ochrochloron Species Due to Exposure to Pyrene," Microbiological Research, Vol. 160, No. 4, 2005, pp. 375-383. http://dx.doi.org/10.1016/j.micres.2005.03.001
75. J. E. L. Corry, "Relationships of Water Activity to Fungal Growth," In: V. N. Reinhold, Ed., Food and Beverage Micology, Beuchat LR, New York, 1987, pp. 51-99.
76. J. Lacey, "Preand Post-Harvest Ecology of Fungi Causing Spolage of Foods and Other Stored Products," Journal of Applied Bacteriology, Symposium Supplement, 1989, pp. 11S-25S.
77. E. Ercoli, J. Gálvez, M. Di Paola, J. Cantero, S. Videla, M. Medaura and J. Bauzá, "Análisis y Evaluación de Pará- metros

Críticos en Biodegradación de Hidrocarburos en Suelo," In: C. Producción III, Ed., Workshop Latinoamericano Sobre Aplicaciones de la Ciencia en la Ingeniería de Petróleo, Formato electrónico, Puerto Iguazú, 2000.
78. V. Sanchis, F. Lafuente, I. Vinas, M. Torres and R. Canela, "Influence of Incubation Conditions in the Patulin Production by Penicillium griseofulvum Dierckx," Rev.

Chapter 4

An Application of the Taguchi Method (Robust Design) to Environmental Engineering: Evaluating Advanced Oxidative Processes in Polyester-Resin Wastewater Treatment

Messias Borges Silva[1,2], Livia Melo Carneiro[1], João Paulo Alves Silva[1], Ivy dos Santos Oliveira[1], Hélcio José Izário Filho[1], and Carlos Roberto de Oliveira Almeida[1]

[1]Department of Chemical Engineering, School of Engineering of Lorena EEL, São Paulo University USP, Lorena, Brazil

[2]Production Department, Universidade Estadual Paulista-Faculdade de Engenharia do campus de Guaratinguetá—UNESP, Guaratinguetá, Brazil

ABSTRACT

This paper presents the Taguchi Method, a statistical design modelling for experiments applied in environmental engineering. This method was applied to optimize the treatment conditions of polyester-resin effluent by means of Advanced Oxidative Processes (AOPs) using chemical oxygen demand (COD) as response parameter. The influence of each independent parameter including respective interactions was evaluated by Taguchi Method, which allowed determining the most statistically significant variables and conditions to best fit the process. Results showed that Taguchi Method is a very useful tool for environmental engineering field and possible simplifications of analysis and calculations through commercially available software.

INTRODUCTION

Genichi Taguchi developed the foundations of robust design introduced in the 1950s and 1960s. As cited by Padke [1] the Taguchi Method may be applied to a wide variety of problems. The application of the method in electronics, automotive products, photography and many others industries had been an important factor in the rapid industrial growth of Japanese industries. The Taguchi Method is based on many ideas extracted from standard statistical design of experiments (SDOE or DOE) [2] - [4], because a conventional design optimization may not always satisfy the desired targets [5]. Furthermore, this method provides other advantages such as economical reduction and variability of the response variable; it also ensures an optimum decision during laboratory experiments or plant operations. It is an important tool to improve the productivity of the research and development activity; it reduces economically the number of trials being applied to any processes, even including environmental engineering and can be used to optimize processes.

In this paper, all steps associated to the application of the Taguchi Method will be described, as well as the tools (orthogonal arrays OA, signal-to-noise ratio S/N and analysis of variance ANOVA) illustrating an application in the environmental engineering field, using advanced oxidative process (AOP) for the treatment of a wastewater from a polyester-resin factory.

Physical and chemical processes have been the most applied processes in such effluents treatments, which include air stripping at different pH values, powdered activated carbon (PAC), adsorption, filtration and EDTA chelation. Respirometry was used for toxicity reduction evaluation after physical and chemical effluent fractionation. From all procedures investigated, only air stripping was significantly effective in reducing wastewater toxicity. Air stripping in pH 7 reduced toxicity in 18.2%, while in pH 11 a toxicity reduction of 62.5% was observed [6].

The removal of diethyleneglycol (DEG) and several polyether-polyols of different molecular weights by ultrafiltration were studied. These polyether-polyols consist of polyethyleneglycols (PEG) of different molecular weights (800 and 6000) and two ethylene oxide-propylene oxide copolymers: Pluronic PE6100 and Alcupol F4811. No significant retention was obtained for DEG and PEG-800, but for PEG-6000, rejection coefficients higher than 80% were reached. Temperature effects, transmembrane pressure and feed rate on both permeate flux and rejection coefficients were also studied. Secondly, the same procedure was followed by 0.1% w/w solutions of Alcupol and Pluronic, obtaining rejection coefficients of almost 100% [7]

Advanced Oxidative Process (AOP) is an important alternative for treatment of contaminated water and wastewater containing hardly-biodegradable anthropogenic substances, pharmaceuticals, pesticides, disinfections of drinking waters [8]-[15] and used after well-established methods (flocculation, precipitation, adsorption, etc.) and before biological methods (aerobic activated sludge) to enhance the biodegradability of wastewater. AOPs modify the pollutants structure producing less toxic and biodegradable products, which can be treated by a biological process. AOPs can be defined as methods where hydroxyl radicals (HOŸ) are produced in sufficient quantities to act as main oxidizing agent. Hydroxyl radicals can be generated as a result of the combination of strong oxidizing agents, such as hydrogen peroxide and ozone. Ultraviolet (UV) or visible radiation and catalysts such as metal ions and semiconductors can also be used to create hydroxyl radicals [16] [17] AOPs using the combination of ozone with other oxidant agents (UV radiation and hydrogen peroxide) allow only oxidation of dissolved organic compounds that are normally refractory to the direct attack of ozone. UV radiation and hydrogen peroxide addition lead to the ozone demolition and hydroxyl radical formation [18].

There are many types of AOPs applied for the oxidation of pollutants in water and wastewater, as catalytic ozonation (O_3) [19] [20], or combination of hydrogen peroxide with ozone (H_2O_2/O_3) or ultra violet (H_2O_2/UV), and UV/titanium dioxide [21] or UV/zinc oxide [22], treating pollutants by oxidation and by hydroxyl radical [23]. Despite AOPs have been widely studied to reduce toxicity of various types of industrial effluents, the use of AOPs in the treatment of effluents from polyester resins facture has been little used.

In this context, the present study aims to apply the Taguchi Method to evaluate advanced oxidative process for the treatment of a recalcitrant effluent from polyester-resin facture.

MATERIALS AND METHODS

Steps of the Taguchi Method

The following steps for implementing Taguchi experimental design according to Barrado et al. [24] are:

1) Select the output variable to be optimized, 2) identify factors (input variables) affecting output variable (response) and choose

levels to be tested, 3) select orthogonal array, 4) assign factors and interactions to the columns of the array, 5) perform experiments, 6) carry out a statistical analysis and the signal-to-noise ratio and determine the optimal conditions to adjust factor levels and 7) perform confirmatory experiment, if necessary.

Planning the Experiment

In this stage, the researcher must define or choose the factors (independent variables or input variables) and the respective operation levels. After this choice, it is necessary to find out the best or economical matrix experiment (orthogonal array OA), which can be obtained in the literature [1] [25] Then, choose the desired signalto-noise ratio function (smaller-the-better, larger-the-better, nominal-the-better), equation 1, equation 2 and equation 3, respectively.

Signal-to-Ratio S/N

The signal-to-noise ratio is a logarithmic function used to optimize the process or product design, minimizing the variability. The signal-to-noise ratio can also be understood as an inverse of variance and maximization of signal-to noise ratio allowing reduction of the variability of the process against undesirable changes in neighbor environment (also named uncontrollable factor or factor of noise). To minimize variability, we must choose the level of factor that produces the maximum value of S/N.

Three types of common problems and respective signal-to-ratio function are presented as follows:

Smaller-the better

$$S/N = -10\log\left[\frac{1}{n}\sum_{i=1}^{n} y_i^2\right] \quad (1)$$

where y_i denote the n observations of response variable Larger-the-better

$$S/N = -10\log\left[\frac{1}{n}\sum_{i=1}^{n}\frac{1}{y_i^2}\right] \qquad (2)$$

Nominal-the-better

$$S/N = 10\log\frac{\mu^2}{\sigma^2} \qquad (3)$$

where μ_i^2 denotes the square of mean and s^2 the variance of the observations of the replicated response values.

Here it is important to emphasize the needs of randomizing the trials to minimize systematic error. To illustrate this paper, a wastewater obtained from a polyester-resin factory was used and the following factors were defined as presented in Table1 An orthogonal array L_{16} was used which means 16-trial experimental matrix, presented in Table2 The orthogonal array L_{16} was chosen by using the degree of freedom method as cited by Padke [1].

Orthogonal Array OA and Linear Graphs

Orthogonal array is a special experimental matrix designed by L_i, where i is the number of trials of experimental matrix or total degree of freedom and consists of a set of experiments where we change the settings of process parameters. The use of OA allows the effects of several process parameters to be determined efficiently, Padke [1] . Here, orthogonality is interpreted in the combinatory sense; that is, for any pair of columns, all combinations of factor levels occur and they occur at an equal number of times. This is called balancing property and it implies orthogonality. Each column of AO has associated one degree of freedom (number of levels minus one) and can be assigned one factor or interaction. A list of orthogonal arrays and linear graphs can be obtained in Padke [1] and Taguchi and Konishi [25].

Linear Graphs

To enhance the flexibility of arrays, Dr. Taguchi used linear graphs to represent the arrays. By using these graphs and triangular tables provided by Taguchi, the experimenter can effectively study interactions between experimental factors as well as effects of individual factors (main effects) themselves. Linear graphs make this possible by providing a logical scheme for assigning interactions to the orthogonal array without confounding effects of interactions with effects of the studied individual factors. In this paper, a linear graph Taguchi L_{16} (Figure 1) was used, which presents an arrangement with seven factors (points 1, 2, 4, 8, 10, 12 and 15) and eight interactions (lines 3, 5, 6, 7, 9, 11, 13 and 14). Factors distribution in the linear graph is made in order to obtain interactions that might prove to be more significant [1].

Table 1: Factors and levels evaluated in Taguchi L_{16} Orthogonal Array

	Factor	Level 1	Level 2
A	pH	3	5
B	Temperature (°C)	25	35
C	Fe^{2+} Concentration (g·l^{-1})	0.6	1.2
D	H$_2$O$_2$ Concentration (g·l^{-1})	14	28
E	Ozone Flow Rate (g·h^{-1})	0.10	0.21
F	Stirring (rpm)	70	150
G	UV radiation	Absent	present

Table 2: Taguchi L_{16} Orthogonal Array and response variables

Assay	A1 pH	C2 Fe	AxC3 pH/Fe	D4 H_2O_2	AxD5 pH/H_2O_2	CxD6 H_2O_2/Fe	ExG7 O_3/UV	E8 O_3	AxE9 pH/O_3	F10 Stir	H11 pH/Stir	B12 T	AxB13 pH/T	AxG14 pH/UV	G15 UV
1	1	1	1	1	1	1	1	1	1	1	1	1	1	1	1
2	1	1	1	1	1	1	1	2	2	2	2	2	2	2	2
3	1	1	1	2	2	2	2	1	1	1	1	2	2	2	2
4	1	1	1	2	2	2	2	2	2	2	2	1	1	1	1
5	1	2	2	1	1	2	2	1	1	2	2	1	1	2	2
6	1	2	2	1	1	2	2	2	2	1	1	2	2	1	1
7	1	2	2	2	2	1	1	1	1	2	2	2	2	1	1
8	1	2	2	2	2	1	1	2	2	1	1	1	1	2	2
9	2	1	2	1	2	1	2	1	2	1	2	1	2	1	2
10	2	1	2	1	2	1	2	2	1	2	1	2	1	2	1
11	2	1	2	2	1	2	1	1	2	1	2	2	1	2	1
12	2	1	2	2	1	2	1	2	1	2	1	1	2	1	2
13	2	2	1	1	2	2	1	1	2	2	1	1	2	2	1
14	2	2	1	1	2	2	1	2	1	1	2	2	1	1	2
15	2	2	1	2	1	1	2	1	2	2	1	2	1	1	2
16	2	2	1	2	1	1	2	2	1	1	2	1	2	2	1

Table 3: Taguchi L_{16} Orthogonal Array and results obtained in the detoxification treatment by Advanced Oxidative Processes

Assays	A1	C2	AC3	D4	AD5	CD6	EG7	E8	AE9	F10	AF11	B12	AB13	AG14	G15	COD Reduction	SD
	pH	Fe^{2+} (g·l^{-1})	pH/ Fe^{2+}	H_2O_2 (g·l^{-1})	pH/ H_2O_2	H_2O/ Fe^{2+}	O_3/ UV	O_3(g·h^{-1})	pH/ O_3	Stirring	pH/ Stirring	Temp	pH/ Temp	pH/ UV	UV		
1	3	0.6	1	14	1	1	1	0.10	1	70	1	25	1	1	absent	6.34	0.11
2	3	0.6	1	14	1	1	1	0.21	2	150	2	35	2	2	present	24.36	0.26
3	3	0.6	1	28	2	2	2	0.10	1	70	1	35	2	2	present	27.99	0.27
4	3	0.6	1	28	2	2	2	0.21	2	150	2	25	1	1	absent	12.16	2.79
5	3	1.2	2	14	1	2	2	0.10	1	150	2	25	1	2	present	19.52	0.62
6	3	1.2	2	14	1	2	2	0.21	2	70	1	35	2	1	absent	18.10	0.98
7	3	1.2	2	28	2	1	1	0.10	1	150	2	35	2	1	absent	17.13	0.26
8	3	1.2	2	28	2	1	1	0.21	2	70	1	25	1	2	present	33.30	0.70
9	5	0.6	2	14	2	1	2	0.10	2	70	1	25	2	1	present	12.97	0.96
10	5	0.6	2	14	2	1	2	0.21	1	150	2	35	1	2	absent	7.49	0.34
11	5	0.6	2	28	1	2	1	0.10	2	70	1	35	1	2	absent	10.28	2.88
12	5	0.6	2	28	1	2	1	0.21	1	150	2	25	2	1	present	23.95	2.16
13	5	1.2	1	14	2	2	1	0.10	2	150	1	25	2	2	absent	10.46	0.09
14	5	1.2	1	14	2	2	1	0.21	1	70	2	35	1	1	present	24.40	4.97
15	5	1.2	1	28	1	1	2	0.10	2	150	1	35	1	1	present	34.09	0.83
16	5	1.2	1	28	1	1	2	0.21	1	70	2	25	2	2	absent	16.39	2.6

Performing the Experiment

By this time, the researcher must conduct the matrix experiment to obtain values of the response variable and the signal-to-noise ratio. After conducting the matrix experiment, the results fill the Table 2 and proceed the calculus of signal-to-noise ratio. Here, we reinforce the need to randomize the trials.

Analyzing the Experimental Results

In this step the researchers fulfill Table 2 and perform the calculus of the effects of factors, signal-to-noise ratio and ANOVA.

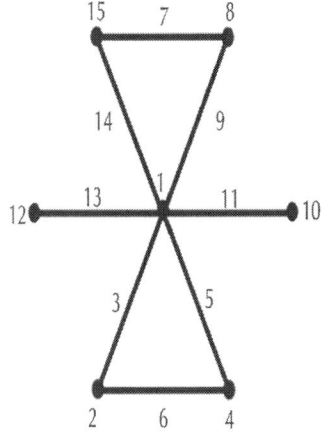

Figure 1: Linear graph Taguchi L_{16}.

Calculus of Effects

Main effects of a factor level are defined as the deviation caused from the overall mean. Using our illustration to evaluate the effect of factor A, shown in Table 2, the mean under condition A_1 assigned by m_{A1} and the mean under condition A_2 assigned by m_{A2} can be calculated by:

$$m_{A1} = \frac{1}{8}(R1 + R2 + R3 + R4 + R5 + R6 + R7 + R8) \quad (4)$$

where Ri is the mean value of response at trial i.

The effect can be interpreted as the mean of responses when factor is set on a specified level.

$$m_{A2} = \frac{1}{8}(R9 + R10 + R11 + R12 + R13 + R14 + R15 + R16) \quad (5)$$

The effects of every other factor are equally calculated.

Analysis of Variance (ANOVA)

ANOVA is used to find out main and interaction effects of categorical independent variables (called "factors") on an interval dependent variable. In that context, ANOVA is used to judge how strong the influence of the factors over the response variable is or how different factors affect the response variable to a different degree. About ANOVA there are many introductory texts on elementary statistical theory books.

Apparatus and Chemicals

The basic aim of this work is to treat a wastewater from polyester and alkyl resin manufacturing industry using the Taguchi Method. The wastewater is a mixture of various organic raw materials unreacted (vegetable oils, abietic acid, benzoic acid, phtalic anidride, hexahydrophtalic anidride, butyl phenol, neopentyl glycol, xylol, toluol, isophtalic acid, terephtalic acid, maleic anidride, penthaerithritol) and polyester. The manufacturing process generates a wastewater with a COD content of over 220,000 mg·l^{-1}. Chemical Oxygen Demand (COD) analysis was carried out by a FEMTO-600 spectrophotometer using dichromate solution as the oxidant in strong acid medium. Color is developed during the oxidation and measured against a water blank using a colorimeter. The change of the dichromate solution color was determined at l = 620 nm.

Polyester-resin effluent was submitted to different treatments conditions by homogeneous AOPs. The treatments were conducted in a glass reactor with a capacity of 400 ml, equipped with a water-cooled system and a magnetic stirring, as shown in Figure 2. Ultraviolet radiation was provided by a medium-pressure mercury vapor lamp (125 W Phillips) placed in the solution through a quartz bulb[26] [27] . A GEHAKA pH meter was used for pH measurement. Ozone was bubbled through the solution using a dielectric barrier discharge Ozonator. Air was bubbled through the solution using a conventional compressor. A magnetic stirrer was used throughout the experiment to ensure a homogeneous medium. Fenton reagent (mixture of H_2O_2 and Fe^{2+} cation) was prepared using H_2O_2 30 % v/v and $FeSO_4$ 0.18 mol·l^{-1}.

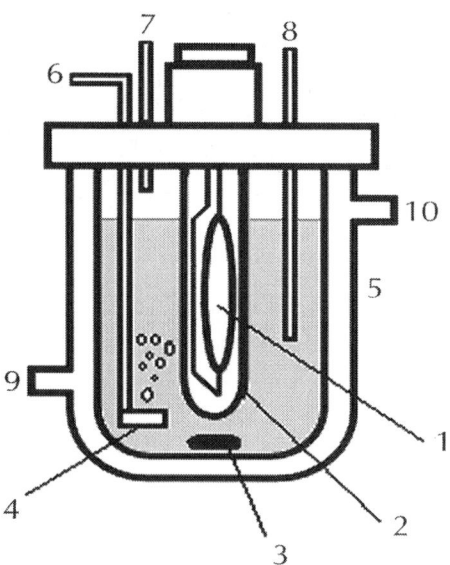

Figure 2: Schematic representation of the conventional photochemical reactor (1-mercury vapor lamp; 2-quartz bulb; 3- magnetic bar; 4-O_3 disperser; 5-cooling jacket; 6-inlet O_3; 7-gas outlet; 8-sample removal.

The percent reduction of COD was conducted by applying a Taguchi experimental design [24] [28]. The choice of factors and their levels (Table 1) were made after brainstorming with people who work on environmental management field. The level of factors was adjusted according to the respective assays (Table 2) to perform the AOPs treatment. After the cell reactor was fulfilled, the electric stirrer, temperature probe, pH electrode, ozone-feeding and air-feeding glass tube were connected and UV generator turned on. The pH was adjusted by using hydrochloric acid or sodium hydroxide solutions.

RESULTS AND DISCUSSION

Aiming to evaluate the potential application of AOPs in reducing the toxicity of polyester resin effluent, assays were conducted to study the influence of Fe (II) concentration, hydrogen peroxide concentration, ozone concentration, UV radiation, pH, temperature and stirring. Chemical Oxygen Demand (COD) was selected as the response variable to be optimized (minimized). To avoid a multivariate analysis and use a minimum set of experiments one Taguchi L_{16} Orthogonal Array was chosen. Taguchi L_{16} design orthogonal array permits to find the effects of each individual controllable factor on the response (% of COD reduction), which are presented in Table 3

- Experimental condition in which we observed largest reduction of COD (33.3%) of the treated effluent was the assay 15 (Table 3), in which the concentration of Fe^{2+} and H_2O_2 (1.2 g·l^{-1} and 28 g·l^{-1}, respectively) were the highest level, indicating a possible positive effect of higher concentrations of these variables on the reduction of COD. It is important to note that the assay was carried out in presence of UV radiation and absence of ozone, and at pH 5. A similar reduction of COD (33%) was observed in assay 8, where experimental conditions are similar to those of assay 15, except for the pH conditions and ozone.

- Assays performed under the presence of UV radiation, COD reduction was above 20%, except for assay 9, which showed 13% COD reduction, even in the presence of UV radiation. However, in this experiment the most relevant parameters were at level 1, which corresponds to minimum concentrations of hydrogen peroxide and Fe^{2+} concentration, and pH 5. Results of COD reduction showed that assays performed in the absence of UV radiation reached COD reduction of about 12%, which means half of the mean value of COD reduction obtained for assays performed in the presence of UV radiation. This result suggests a major importance of the UV radiation presence on the polyester resin effluent degradation.
- In general, the largest reductions of COD were obtained when the H_2O_2 concentration factor was at the highest level and in the presence of UV radiation, in assays 3, 8, 12 and 15. The greatest COD reductions were observed in assays 8 and 15 (33% and 34%, respectively), which were carried out at Fe^{2+} concentration at its highest level, even in different conditions of pH and ozone concentration. Ph is a significant parameter on the pollutant AOPs degradation, for example, in Fenton and photo-Fenton processes, many studies have reported pH values around three [16] . Generally, at pH values above four, a decrease on hydrogen peroxide decomposition rate occurs due Fe^{2+} species precipitation as iron hydroxide [16] . However, in this study pH presented a low effect on the degradation process within the pH range studied.

Statistical Analysis

The statistical significance of main effects and interactions on the reduction in COD for the polyester resin effluent treatment was confirmed by analysis of variance effects (ANOVA) (Table 4). Statistical significance of the factors effects was considered at a 95% confidence level and it is related by p-value in the ANOVA. Ac

SD = standard deviation; COD = chemical oxygen demand.

Table 4: Analysis of variance for the response evaluated in the experimental design

Variation source	Sum of squares	DF	Mean sum of squares	F	p		Contribution (%)
pH	13.3416	1	13.3416	6.1950	0.024205	*	2.1
Fe^{2+}	94.6594	1	94.6594	43.9539	0.000006	*	15.0
pH/Fe^{2+}	1.7633	1	1.7633	0.8188	0.378960		0.3
H_2O_2	80.6292	1	80.6292	37.4392	0.000015	*	12.8
pH/H_2O_2	1.7887	1	1.7887	0.8306	0.375632		0.3
H_2O_2/Fe^{2+}	1.3661	1	1.3661	0.6343	0.437435		0.2
O_3/UV	0.3172	1	0.3172	0.1473	0.706190		0.1
O_3	20.8645	1	20.8645	9.6882	0.006701	*	3.3
pH/O_3	4.0830	1	4.0830	1.8959	0.187498		0.6
Stirring	0.2190	1	0.2190	0.1017	0.753909		0.0
pH/Stirring	1.2061	1	1.2061	0.5600	0.465111		0.2
Temperature	24.3201	1	24.3201	11.2927	0.003979	*	3.9
pH/Temperature	16.5279	1	16.5279	7.6745	0.013652		2.6
pH/UV	0.0286	1	0.0286	0.0133	0.909634		0.0
UV	333.7682	1	333.7682	154.9811	0.000000	*	53.0
Residue	34.4577	16	2.1536				
Total	629.3408	31					R^2= 94.5

DF = degree of freedom cording to ANOVA (design L_{16}), significant variables for polyester resin effluent treatment were Fe^{2+} concentration, hydrogen peroxide and ozone concentration, and pH, temperature and UV radiation. Among these variables, Fe^{2+} and H_2O_2 concentration, and UV radiation were considered significant with a confidence level of 99%. The significant variable was responsible for 15, 12.8 and 53% for Fe^{2+} concentration, H_2O_2 concentration and UV radiation, respectively, of all the observed variation (Table 4). The analysis of variance indicates that, in the experimental evaluated range, photo-Fenton (utilization of Fe^{2+} and H_2O_2 in the presence of UV radiation) was the most effective method for the polyester-resin effluent treatment.

The signal-to-noise (S/N) ratio concept was used to evaluate the variability and how it changes around the mean, by influence of external uncontrollable factor (noises). In this work, the goal is to reduce COD, e.g, smaller COD is better. In that condition the response is optimized and the process will be robust against external environment or uncontrollable factors [1]. As an optimized parameter COD is a smaller-is-better, the calculation of S/N was made using the Equation (1). Note that L_{16} orthogonal array needs 16 assays for 6 factors operating at 2 levels. That experimental matrix allows studying many factors, saving costs and time if compared with traditional full factorial design or conventional experiment in which one factor has the level changed while the other factors keep constant.

In signal-to-noise graph (Figure 3), it may be observed the effects of the variables (A-pH; B-Temperature; C-Fe^{2+}; D-H_2O_2; E-O_3; F-Stirring; G-UV) and its interactions (AC; AD; CD; EG; AE; AF; AB; AG) studied in Taguchi L_{16} Orthogonal Array on the response COD reduction. Similar to the analysis of variance the signal-tonoise analysis showed that the significant factors were Fe^{2+} and H_2O_2 concentration, and UV radiation, and all of others variables had a greater effect when set at the highest level. Through the signal-to-noise ratio graph it was observed that the ozone and temperature factors were better for the treatment process when used on its highest level.

Therefore, based on the most significant factors (concentration of H_2O_2 and Fe^{2+}, and UV radiation), we can conclude that the most appropriate treatment for polyester resin effluent is the photo-Fenton process. pH, ozone, temperature and agitation factors, although they are not significant at statistical analysis, they are important factors which should be kept at the level of best fit (pH 3, ozone flow rate of 0.21 g·h^{-1}, temperature of 35°C and agitation of 150 rpm).

Simplification of Calculus and Analysis

All steps of the Taguchi Methods can be simplified by using a statistical software. Nowadays, there are several types of commercial software with Taguchi's options. These are the most recommendable: MINITAB, STATISTICA, STATGRAPHICS, DESIGN EXPERT, QUALITEK and others. In this work, the STATISTICA 6.0 software was used.

CONCLUSIONS

The Taguchi Method can facilitate and optimize the work of experimenters. This method is an important means of addressing several variables simultaneously (multivariate analysis), with fewer trials, lower costs, and a shortened duration.

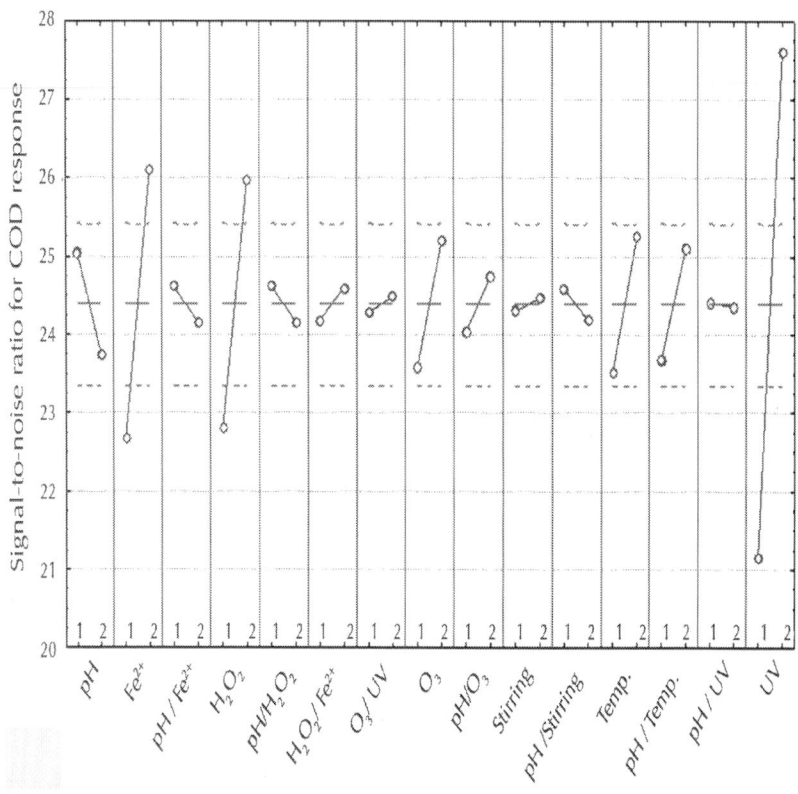

Figure 3: Effects of the factors (A-pH; B-Temperature; C-Fe^{2+}; D-H_2O_2; E-O_3; F-Stirring; G-UV) and its interactions (AC; AD; CD; EG; AE; AF; AB; AG) evaluated in Taguchi L_{16} Orthogonal Array for COD reduction of the polyester-resin treated by Advanced Oxidative Processes. Level 1: lower; level 2: higher.

Initial concentrations of Fe^{2+}, hydrogen peroxide and ultraviolet radiation were statistically revealed to be the most susceptible to influence utilizing the Taguchi Method. The photo-Fenton process was the most effective in reducing the COD of polyester-resin effluent. Other factors that had less influence within the range of variation studied (pH, temperature, agitation, and ozone concentration), should result in a maintained best-fit level as defined in the Taguchi experimental design. The treatment by means of advanced oxidative process provided an approximate 35% reduction in chemical oxygen demand of the polyester-industry wastewater. However, when compared to studies describing the treatment of this effluent by Advanced Oxidative Processes, we see that the results were relevant. This demonstrates that the Taguchi Method is a very useful tool for applications in the fields of environmental engineering and industrial wastewater treatment.

ACKNOWLEDGMENTS

The authors gratefully acknowledge financial support from Coordenação de Aperfeiçoamento de Pessoal de Nível Superior (CAPES), and Conselho Nacional de Desenvolvimento Científico e tecnológico (CNPq)-BRAZIL.

REFERENCES

1. Padke, M.S. (1989) Quality Engineering Using Robust Design. Prentice Hall, Englewood Cliffs.
2. Dawson, E.A. and Barnes, P.A. (1992) A New Approach to the Statistical Optimization of Catalyst Preparation. Applied Catalysis A: General, 90, 217-231. http://dx.doi.org/10.1016/0926-860X(92)85060-O
3. Robin, A., Alves de Souza, K., Rosa, J.L. and Silva, M.B. (2002) Electrodeposition of Copper as a Route for Tantalum Drawing. Surface Engineering, 18, 120-125.
4. Rosa, J.L., Robina, A., Silva, M.B., Baldana, C.A. and Peres, M.P. (2008) Electrodeposition of Copper on Titanium Wires: Taguchi Experimental Design Approach. Journal of Materials Processing Technology, 209, 1181-1188. http://dx.doi.org/10.1016/j.jmatprotec.2008.03.021
5. Zang, C., Friswell, M.I. and Mottershead, J.E. (2005) A Review of Robust Optimal Design and Its Application in Dynamics. Computers & Structures, 83, 315-326. http://dx.doi.org/10.1016/j.compstruc.2004.10.007

6. Caffaro-Filho, R.A., Morita, D.M., Wagner, R. and Durrant, L.R. (2009) Toxicity-Directed Approach of Polyester Manufacturing Industry Wastewater Provides Useful Information for Conducting Treatability Studies. Journal of Hazards Material, 163, 92-97. http://dx.doi.org/10.1016/j.jhazmat.2008.06.063
7. Canizares, P., Pérez, A., Camarillo, R. and Llanos, J. (2007) Removal of Polyether-Polyols by Means of Ultrafiltration. Desalination, 206, 594-601. http://dx.doi.org/10.1016/j.desal.2006.03.582
8. Oguz, E., Keskinler, B., Celik, C. and Celik, Z. (2006) Determination of the Optimum Conditions in the Removal of Bomaplex Red CR-L Dye from the Textile Wastewater Using O_3, H_2O_2, HCO_3^- and PAC. Journal of Hazardous Materials, 131, 66-72.
9. Meric, S., Kabdalia, I., Tunay, O. and Orhon, D. (1999) Treatability of Strong Wastewaters from Polyesters Manufacturing Industry. Water Science and Technology, 39, 1-7. http://dx.doi.org/10.1016/S0273-1223(99)00247-4
10. Parra, S.P.C. (2001) Coupling of Photocatalytic and Biological Process as a Contribution to the Detoxification of Water: Catalytic and Technological Aspects. Ph.D. Thesis, ècole Polytechnique Fédérale de Lausanne.
11. Legrini, O., Oliveros, E. and Braun, A.M. (1993) Photochemical Process for Water-Treatment. Chemical Reviews, 93, 671-698. http://dx.doi.org/10.1021/cr00018a003
12. Gunten, U., Huber, M.M., Canonica, S. and Park, G.Y. (2003) Oxidations of Farmaceuticals during Ozonation an Advanced Oxidation Processes. Environmental Science & Technology, 37, 1016-1024. http://dx.doi.org/10.1021/es025896h
13. Chiron, S., Alba, A.F., Rodrigues, A. and Calvo, E.G. (2000) Pesticide Chemical Oxidation: State-of-the-Art. Water Resource, 2, 366-377.
14. Silva, M.B., Almeida, C.R.O., Chaves, F.J.M., Izário Filho, H.J. and Mattos, J.R.A. (2003) Treatment of Strong Wastewater Using Advanced Oxidation Process (AOP) and Biological

Process (BP) to Reduction of Chemical Oxygen Demand (COD) in Samples from Polyester Manufacturing Industry. Conferencia Científica Internacional Medio Ambiente Siglo XXI (MAS III), Santa Clara.

15. Guimaraes, O.L.C. and Silva, M.B. (2007) Hybrid Neural Model for Decoloration by UV/H2O2 Involving Process Variables and Structural Parameters Characteristics to Azo Dyes. Chemical Engineering and Processing, 46, 45-51.http://dx.doi.org/10.1016/j.cep.2006.04.005

16. Nogueira, R.F.P., Trovó, A.G., Silva, M.R.A., Villa, R.D. and Oliveira, M.C.O. (2007) Fundamentos e aplicacoes ambientais dos processos Fenton e foto-Fenton. Química Nova, 30, 400-408.http://dx.doi.org/10.1590/S0100-40422007000200030

17. Freire, R.S., Pelegrini, R., Kubota, L.T. and Durán, N. (2000) Novas tendências para o tratamento de resíduos industriais contendo espécies organocloradas. Química Nova, 23, 504-511.http://dx.doi.org/10.1590/S0100-40422000000400013

18. Guzzella, L., Feretti, D. and Monarca, S. (2002) Advanced Oxidation and Adsorption Technologies for Organic Micropollutant Removal from Lake Water Used as Drinking-Water Supply. Water Research, 36, 4307-4318.http://dx.doi.org/10.1016/S0043-1354(02)00145-8

19. Peixoto, A.L.C. and Izário Filho, H.J. (2010) Statistical Evaluation of Mature Landfill Leachate Treatment by Homogeneous Catalytic Ozonation. Brazilian Journal of Chemical Engineering, 27, 531-534.http://dx.doi.org/10.1590/S0104-66322010000400004

20. Peixoto, A.L.C., Silva, M.B. and Izário Filho, H.J. (2009) Leachate Treatment Process at a Municipal Stabilized Landfill by Catalytic Ozonation: An Exploratory Study from Taguchi Orthogonal Array. Brazilian Journal of Chemical Engineering, 26, 481-492.http://dx.doi.org/10.1590/S0104-66322009000300004

21. Salazar, R.F.S. and Izário Filho, H.J. (2010) Aplicacao de processo oxidativo avancado (POA) como pré-tratamento

de efluente de laticínio para posterior tratamento biológico. Analytica (Sao Paulo), 45, 60-61.
22. Samanamud, G.R.L., Lourea, C.C.A., Souza, A.L, Salazar, R.F.S., Oliveira, I.S., Silva, M.B. and Izário Filho, H.J. (2012) Heterogeneous Photocatalytic Degradation of Dairy Wastewater Using Immobilized ZnO. ISRN Chemical Engineering, 2012, Article ID: 275371. http://dx.doi.org/10.5402/2012/275371
23. Fahami, Nishijima, W. and Okada, M. (2003) Improvement of DOC Removal by Multi-Stage AOP-Biological Treatment. Chemosphere, 50, 1043-1048.http://dx.doi.org/10.1016/S0045-6535(02)00617-3
24. Barrado, E., Vega, M., Grande, P. and Del Valle, J.L. (1996) Optimization of a Purification Method for Metal-Containing Wastewater by Use of a Taguchi Experimental Design. Water Resource, 30, 2309-2314.
25. Taguchi, G. and Konishi, S. (1987) Taguchi Methods Orthogonal Arrays and Linear Graphs: Tools for Quality Engineering. American Supplier Institute, Dearborn, Michigan.
26. Lizama, C., Freer, J., Baeza, J. and Mansilla, H.D. (2002) Optimized Photodegradation of Reactive Blue 19 on TiO2 and ZnO Suspensions. Catalysis Today, 76, 235-246.http://dx.doi.org/10.1016/S0920-5861(02)00222-5
27. Rodríguez, M., Abderrazik, N.B., Contreras, S., Chamarro, E., Gimenez, J. and Esplugas, S. (2002) Iron(III) Photoxidation of Organic Compounds in Aqueous Solutions. Applied Catalysis B: Environmental, 37, 131-137.http://dx.doi.org/10.1016/S0926-3373(01)00333-2
28. Silva, M.B. (1996) Estudo das Condicoes de Preparacao, Caracterizacao e Reatividade de Catalisadores de Prata Suportada em Alumina. Tese de Doutorado, Faculdade de Engenharia Química/Universidade Estadual de Campinas—UNICAMP.

Chapter 5

Effect of Two Liquid Phases on the Separation Efficiency of Distillation Columns

Gardênia Marinho Cordeiro, Stephanie Rolim Dantas, Luís Gonzaga Sales Vasconcelos, and Romildo Pereira Brito

Department of Chemical Engineering, Federal University of Campina Grande, Campina Grande, Brazil

ABSTRACT

Distillation is one of the oldest and most important separation processes used in the chemical and petrochemical industries. On the other hand, it is a process the thermodynamic efficiency of which is very low, and therefore reducing the consumption of energy is one of the targets of research studies on distillation.

This article arose from seeking to reduce energy consumption in a distillation train of 1, 2-dichloroethane (ethylene dichloride-EDC) of a commercial plant producing vinyl monochloride (VMC), which involves an azeotropic distillation column. The reduction in the reboiler heat duty caused significant changes in concentration and temperature profiles throughout the column due to the formation of two liquid phases. The results show that, although very small in percentage terms (less than 2.5%), the appearance of the 2nd liquid phase causes significant changes in the operation of the column and the separation achieved.

INTRODUCTION

Distillation is one of the oldest and most important separation processes used in chemical processes. On the other hand, its thermodynamic efficiency is extremely low, which accounts for the high percentage of global energy consumed in a plant. In general, distillation column reboilers consume over 50% of the energy involved in the process of heat exchange in a plant (Soave and Feliu, 2002 [1]).

The term azeotropic distillation is applied to the class of techniques based on fractional distillation in which azeotropic behavior is exploited to achieve separation. Traditionally, the specie that causes the azeotropic behavior is added as a mass separating agent: the entrainer. In some situations it may be present in the feed mixture (self-entraining) of the azeotropic column (Perry et al., 1999 [2]).

Although a large number of studies involve azeotropic distillation, most involve columns in which a third component is added in order to further the separation. Such studies are about choosing the third component, the influence of a thermodynamic model, evaluating the existence of multiple steady states and the study of process control (Laroche et al., 1992 [3]; Bekiaris et al., 2000 [4]; Magnussen et al., 1979 [5]; Rovaglio and Doherty, 1990 [6]; Wang et al., 1997 [7]; Luyben, 2008 [8]; Wu and Chien, 2009

[9]). Another striking feature of the articles cited is that they consider the formation of two liquid phases only in the reflux vessel.

Lao and Taylor (1994) [10] reviewed the literature on the separation efficiency of distillation columns involving three-phase systems, and cite several sources which give rise to their finding that the conclusions drawn on these systems are contradictory. Some studies claim that overall efficiency was not influenced by the number of liquid phases present. Other studies indicate that the introduction of a second liquid phase may have a strong (positive or negative) influence on the mass transfer.

Widagdo and Seider (1996) [11] published one of the most complete (and even to this day, one of the most cited) articles on the azeotropic distillation process. They showed that knowledge contained in the literature is scant both as to a real understanding of the process and the difficulties regarding control of azeotropic columns. They also emphasized the issue of the formation of two liquid phases within the column, but there is no consensus on the efficiency of separation when columns operating with one and with two liquid phases are compared.

In 1997 Wang et al. [7] observed experimentally the formation of two liquid phases inside a column, depending on the reflux and the reboiler heat duty, as well as the presence of multiple steady states; the study evaluated the dehydration of isopropanol, using cyclohexane as the entrainer.

According to Higler et al. (2004) [12], azeotropic distillation is characterized by its operational complexity, due to the possible formation of two liquid phases inside the column. The authors used an equilibrium and a non-equilibrium stage model and claimed the formation of two liquid phases in the distillation column influences the mass transfer process, thus affecting efficiency.

The equilibrium stage model, widely used in modeling and simulating distillation processes, does not represent the reality that few stages actually operate in equilibrium. This problem can be solved by introducing Murphree efficiencies. However, some authors (Cairns and Furzer, 1990 [13]) warned against incorporating

Murphree efficiencies into equilibrium stage models of three-phase systems. In fact, the projections may be more accurate if a non-equilibrium stage model is considered. However, calculations are complex, thus requiring more computational time, which is not desirable for control applications. But, the biggest obstacle is that the parameters required to perform the calculations are rarely available.

Junqueira et al. (2009) [14] analyzed the formation of two liquid phases in the azeotropic column in the production of anhydrous ethanol, and, in order to decrease this phenomenon, many process configurations have been studied as well as variations in operating conditions. It was concluded that the formation of the second liquid phase may affect the performance of the column and consequently reduce its efficiency. Silva et al. (2003) [15] evaluated the dynamics of an azeotropic distillation column similar to the one considered in this article; however, the entrainer was already present in the feed, which was held in the intermediate region of the column, and the formation of two liquid phase occurred only in the reflux vessel. Guedes et al. (2007) [16] followed the same procedure as the one studied in this paper and, in the steady state, evaluated the process sensitivity relative to the feed temperature; and, dynamically evaluated the influence in feed temperature, including the operation condition with two liquid phases in some plates.

The distillation column considered in this article shows characteristics of an azeotropic distillation, since two liquid phases form in the reflux vessel and, depending on the operation condition, in some stages throughout the column. However, another feature makes the system unconventional: the feed takes place in the reflux vessel. In the research literature few studies have considered systems with these characteristics.

PROBLEM STATEMENT

The distillation column considered in this study is part of the purification train of 1, 2-dichloroethane (ethylene dichloride-EDC) of a commercial plant which produces vinlyl chloride monomer (VCM).

The process of obtaining EDC occurs through the direct chlorination of ethylene (C_2H_4), as shown in the reaction: $C_2H_4 + Cl_2 \rightarrow C_2H_4Cl_2$. The EDC product (high purity) leaves the reactor and moves on to the purification system, where it undergoes aqueous washing. Figure 1 shows the flow diagram of the EDC dehydration process, where it can be observed that aqueous washing is conducted in the separating vessel (or reflux vessel). After the top condenser and in the reflux vessel, there are two liquid phases: an organic one, saturated in H_2O, and an aqueous one, saturated in organic matter. The organic phase returns to the reflux of the column, while the stream of the aqueous phase is discarded.

Although less volatile than the EDC, the H_2O leaves from the top of the column due to the reversal in the value of the constant K (Figure 2), which is due to the fact that H_2O forms a minimal azeotrope, not only with the EDC, but with almost all organic compounds present in the process.

Note that in the stream coming from the reactor (FROMR1) there is no H_2O, so that during washing, the stream that carries out the reflux of the column (TODRY 2) becomes saturated in H_2O.

A close analysis of Figure 1 leads to the conclusion that the system as a whole can be seen as a conventional column (with reboiler, condenser and reflux vessel), with the feed (FROMR1 and WATER) in the reflux vessel. In industry, although the analysis of the degree of freedom indicates two variables will be manipulated, only the reboiler heat duty is used, since the reflux flow rate is used to control the level (organic phase) of the vessel and the distillate flow rate (WASTE) cannot return to the process.

The study by Guedes et al. (2007) [16] aimed at reducing the consumption of energy in the azeotropic column. The question to be answered was: if the reboiler heat duty is the only manipulated variable used, to what extent can it be reduced without compromising the quality of the bottom product (the H_2O mass fraction)? Accordingly, by performing tests in the plant, the reboiler heat duty was gradually reduced, which resulted in plate temperatures (top, middle and bottom) that were much smaller than those observed historically, being indicated. In spite of the amount of moisture in the bottom stream being below the specification (10 ppm), the tests were discontinued after 7 hours of operation, and a new operating condition (lower heat duty) was established.

According to Guedes et al. (2007) [16], a more significant change in the temperature profile occurs because of the formation of a 2nd liquid (aqueous) phase in the plates of the column. And, the good agreement between azeotropic data (Azeotropic Data, 1973 [17]) and solubility (Dechema, 1990) found in the literature for EDC-H_2O and those predicted by the simulations, are the mainstays of this conclusion. However, the simulations were carried out without formally defining an objective function and constraints (optimization). Furthermore, no evaluation of the effect of the possible presence of a 2nd liquid phase in separation was performed. Thus, this study aimed to: formalize optimizing the consumption of energy, and evaluate the efficiency of separation taking two operating conditions into account: without the formation of two liquid phases (Case I) and with the formation of two liquid phases (Case II).

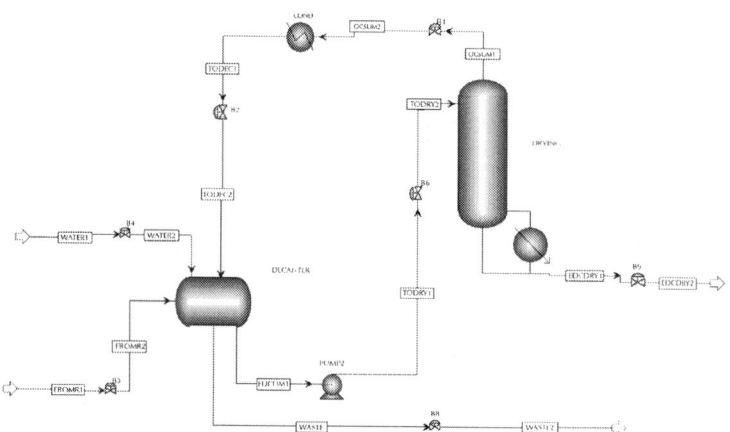

Figure 1: Flow sheet of the EDC dehydration process.

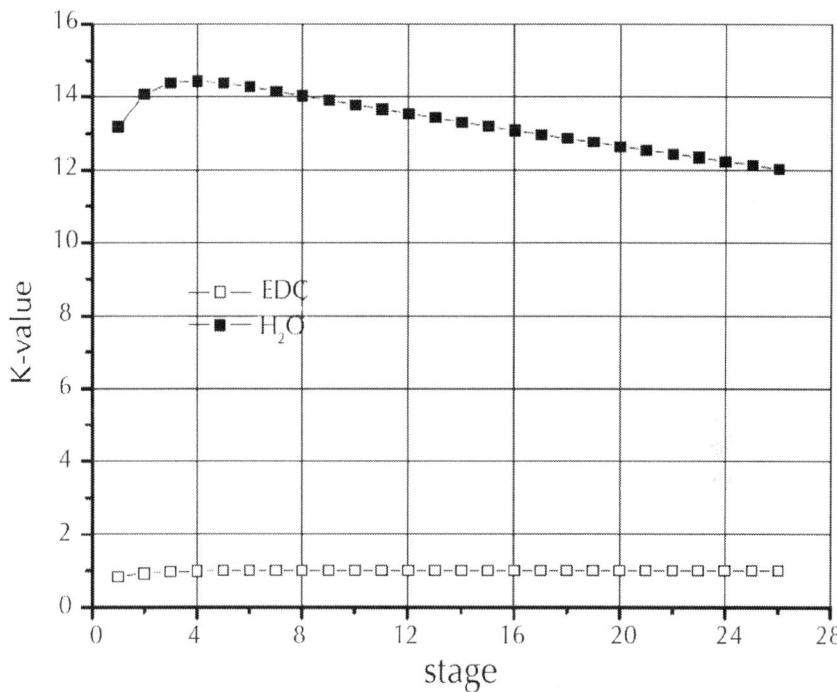

Figure 2: K-values along a column working with a single liquid phase.

MODELING AND SIMULATION

The simulation was performed using Aspen Plus™ commercial simulator. In order to represent the real system, the system was modeled using reboiled absorption, followed by a condenser (Heater) and a decanter (Decanter). To model the column in question, the RadFrac™ routine was used.

The RadFrac™ routine detects the possible formation of a second liquid phase (the main component was H_2O) at any stage; assumes there is an equilibrium stage model; and uses specified values for stage efficiencies. These efficiencies can be manipulated to adapt to the plant data. In this study, a Murphree efficiency equal to 64% for all plates and 100% for the reboiler was used. In the industrial plant, the column has 25 stages (numbered from top to bottom) and a reboiler type thermosyphon. In the Aspen Plus™ simulator, the pressure in each plate of the column, as well as in the other equipment, is kept constant.

To represent the equilibrium between liquid-liquidvapor phases (ELLV), a γ–ϕ procedure was used. Even with the column operating under low pressure, the vapor phase was represented by the Redlich-Kwong Equation of State (EOS). The activity coefficient γ was determined from the NRTL model (Perry et al., 1999 [2]), which represents the ELLV system effectively. Tables 1 and 2, respectively, show the comparison between the azeotropic (Azeotropic Data, 1973) and solubility data (Dechema, 1990 [18]) found in the literature for the EDC-H_2O system (main components) and those predicted by the simulations.

In order to determine the optimal energy consumption, the objective function (J) to be minimized was defined as the reboiler heat duty (Qr).

The restriction in the case of optimization without the presence of two liquid phases (Case I) is the mass fraction of H_2O in the liquid phase (global): if it was not desired to form two liquid phases over the column, the restriction imposed was 2500 Parts Per Million (ppm) (approximately the saturation value of EDC with H_2O at 45°C) for the first stage (numbered from top to bottom) of the column. The choice of this plate was due to its being found that the formation of two liquid phases starts in this plate.

For the operation with two liquid phases (Case II), the restriction imposed was 10 ppm in the bottom stream of the column (the maximum permitted in the plant). Mathematically, the problem was formulated as follows:

Min $J = Qr$ (1)

Subject to

$x_{\#1}^{H_2O} \leq 0.0025$ (2)

Or

$$x_{Bott}^{H_2O} \leq 0.00001 \tag{3}$$

The optimization procedure considered the distillate flow rate (stream OCSUM1) as the manipulated variable (OCSUM1). The objective function was inserted via the Analysis/Optimization Model of the Aspen Plus™ tool, which uses the Sequential Quadratic Programming (SQP) search method for the optimum. The restrictions were inserted using the Analysis/Constraint Model.

Table 1: Comparison of azeotropic data for EDC (1)-H_2O (2) system

Azeotropic boiling Point (1 atm), °C		Mass Fraction of H_2O	
Literature	Aspen Plus™	Literature	Aspen Plus™
71.85	73.85	9.2	9.6

Table 2: Solubility (% weight) of EDC (1)-H_2O (2) system

Temperature, °C (1) in (2)	Literature		Aspen Plus™	
	(2) in (1)	(1) in (2)	(2) in (1)	
30	0.889	0.163	0.888	0.163
40	0.948	0.213	0.940	0.210
50	1.040	0.286	1.023	0.279
60	1.170	0.391	1.149	0.379
70	1.337	0.529	1.331	0.526

The procedure can be implemented over the following steps:
- Fix the number of stages of the column;
- Specify the value of the distillate flow rate, which will be used as an initial estimate;
- Insert, via the Analysis/Optimization Model, the objective function and the range over which the variable may be manipulated;
- Insert, via the Analysis Constraint Model, the restriction and its tolerance.

STEADY-STATE RESULTS

A comparison of data from the plant (the historical operating conditions) and those provided by the simulation is shown in Table 3. The good agreement between real and simulated data, in fact, proves the effectiveness of the modeling and the simulation.

Table 4 shows the conditions of the stream from the reactor (FROMR1) and Table 5 presents results for two operating conditions: 1) historical and 2) optimized.

As per Table 5, with the formation of two liquid phases (Case II), the reduction in energy consumption compared with the situation with a single liquid phase (Case I) is 19.4%; a result caused by a decrease in the distillate flow rate.

The final value of the reboiler heat duty was derived and determined after the constraints were optimized. In both cases, the production of "dry" EDC (EDCDRY2) was very similar.

In Figure 3, note the large difference between the temperature profiles for the two optimal situations. For Case I, a significant variation occurs between the 1st and the 5th plate, and then the rate of increase is almost linear from there to the 26th plate (bottom). On the other hand, in case II, the variation in the rate of increase between the 1st and 16th plate is almost linear, and then there are steep increases in this rate until the 24th plate at which point the temperatures in the two cases coincide.

Table 3: Comparison between the real and simulation data (Guedes et al., 2007)

Variable	Real	Simulation
Reboiler heat duty (kcal/h)	1.52×10^6	1.53×10^6
Temperature at top (°C)	79.0	79.4
Temperature of plate 6 (°C)	85.0	87.0
Temperature at bottom (°C)	93.0	93.4

Table 4: Characteristics of the feed (FROMR1)

Value	
Temperature, °C	40.0
Flowrate, Kg/h	59,250
Mass fraction	
1,1-dichloroethane	0.00009
Carbon-tetrachloride	0.00002
1,2-dichloroethane (EDC)	0.99398
Water	0.00000
1,1,2-trichloroethane	0.00130
1,2,3-trichlorobenzene	0.00461

Table 5: Results for two operational conditions

	Historical	Optimized	
		Case I	Case II
Distillate flow rate (kg/h)	4850.0	4616.9	1465.3
Reboiler heat duty (kcal/h)	1.52×10^6	1.4985×10^6	1.2079×10^6

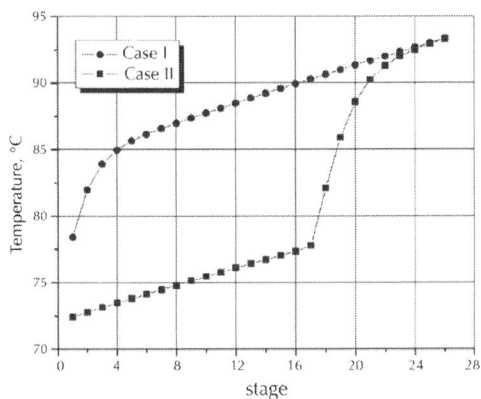

Figure 3: Temperature profiles for the two optimized situations.

In both cases, the linear behavior of the temperature takes place basically by varying the pressure, since the change in the composition of the species along the column is very small, as shown in Figure 4. Simulations that include a negligible pressure drop along the column show the temperature profiles then remain on plateaus, rather than go straight upward, thus confirming this observation on the result of there being negligible drops in the pressure. The profiles obtained experimentally by Wang et al. (1997) [7] show qualitative forms similar to Figure 3. However, unlike the findings of this study, the percentage of H_2O present in the feed was high.

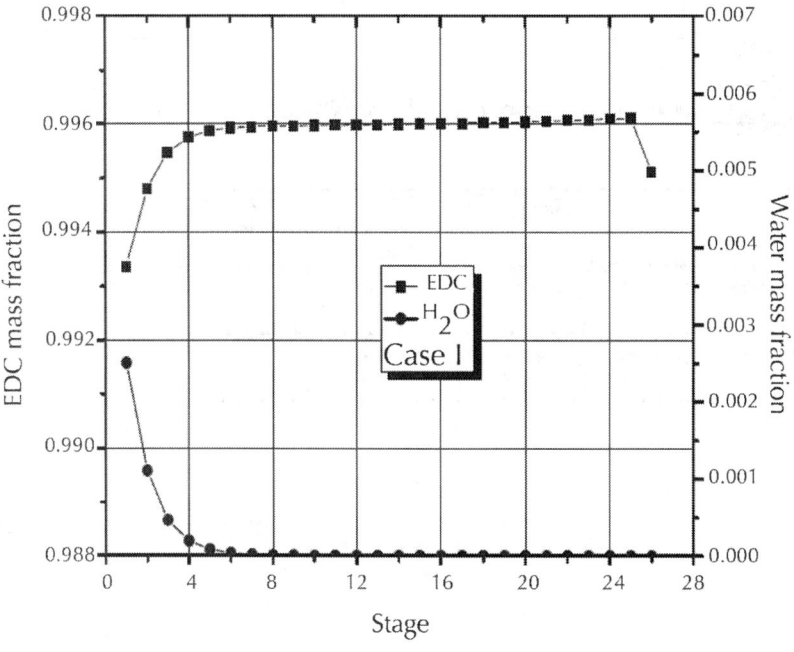

(a)

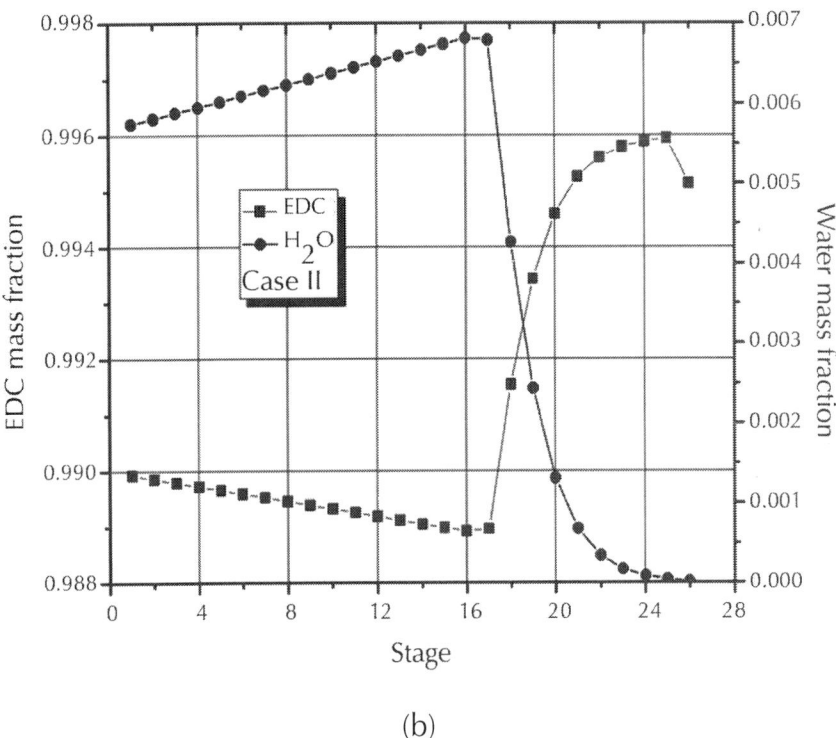

(b)

Figure 4: (a) Composition profile (EDC and H_2O) In the liquid phase (global) for Case I; (b) Composition profile (EDC and H_2O) in the liquid phase (global) for Case II.

Figure 4 shows the mass fraction of EDC and H_2O (main components) in each stage, from which it may be seen that, in each case, the mass transfer is at its most significant in different regions of the column: for Case I in the upper region; for Case II, in the lower one. For Case II, the greatest change in composition occurs in the region where the 2nd liquid phase is not present (from the 16th stage on). In fact, in both cases, dehydration mainly occurs in a small region of the column.

Given the low transfer of mass in most of the column, Figure 4 suggests that the number of stages of the column could be smaller. In fact, if the reboiler heat duty is maintained constant, simulations for a column with 19 stages show the presence of a single liquid phase and the fraction of H_2O at the bottom is within specification. However, for columns with 18 stages, two phases are present and the liquid fraction of H_2O at the bottom (1000 ppm) is above the one laid down in the specification.

The reason for the formation of two liquid phases can be seen in Figure 4. For Case II, in the region of two liquid phases, the maximum mass fraction of H_2O is about 0.7% by weight, so it is above the saturation value of the organic phase with H_2O. For Case I, the maximum mass fraction of H_2O is around 0.25% by weight (approximately the saturation value of EDC with H_2O). The behavior of Case II is due to the fact that the decrease in the reboiler heat duty does not prompt the removal of H_2O (in the form of azeotrope) in the early stages of the column.

Figure 5 shows the Separation Factor (SF) defined by Equation (4) Perry et al., 1999 [2]) along the column, in which what can be noted is that the separation efficiency is increased when there is a single liquid phase. Even if the second liquid phase is present, the Separation Factor is greater in stages where this phase disappears. From this Figure, note also that, for Case II (a two liquid phase up to plate 16), dehydration occurs in the last few plates. Overall, the magnitude of the Separation Factor measured for Case I (1.15E9) was completely different from that calculated for Case II (235).

$$\frac{y_{H_2O}/x_{H_2O}}{y_{EDC}/x_{EDC}} \tag{4}$$

The reduction in the SF for Case II may be explained as a direct consequence of the reduction of the reflux flow rate (caused by the decreased flow of distillate), which is usually one of the variables that most impact separations. However, what needs to be taken into account is that a simulation condition which operates immediately before the 2nd liquid phase forms and which involves a minimal reduction in the reflux flow rate, results in an SF of 1.7E9, that is, in the same order of magnitude of that calculated for Case I. This result is in accordance with various citations in the article by Widagdo and Seider (1996) [11] and as pointed out by Junqueira et al. (2009) [15]. That is to say there is a drastic reduction in the separation efficiency of columns operating with two liquid phases in some plates.

The results presented in Figure 5 were obtained after optimizing the reboiler heat duty and assuming a constant Murphree efficiency (64%). Figure 6 shows the global Separation Factor H_2O/EDC depending on the Murphree efficiency, without considering the optimization. For operation with a single liquid phase (Case I) a distillate flow rate was set at 3500 kg/h, while the condition for the two liquid phases (Case II), this was set at 1450 kg/h.

From the results of Figure 6, it is possible to note that the operation with a single liquid phase in the plates is much more dependent on the operational efficiency in which some of the column plates have two liquid phases. For Case I, the behavior is similar to that typically observed for distillation columns: separation is directly proportional to the efficiency of the stages. Moreover, where two liquid phases are observed in some plates of the column (Case II), separation decreases when the efficiency of the plate is increased, which is caused by increasing the number of plates with two liquid phases (16 to 18).

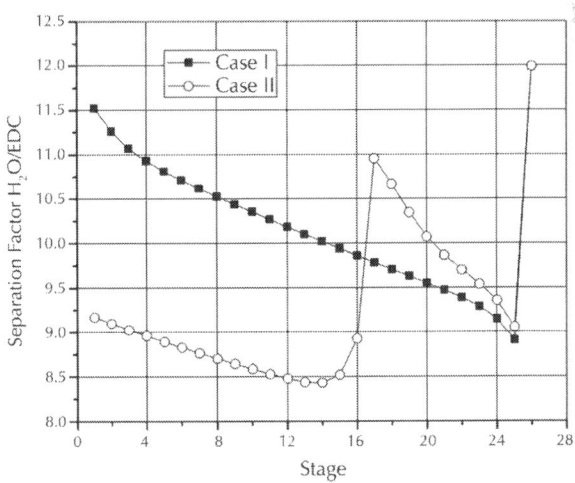

Figure 5: Separation factor H_2O/EDC along the column for two optimized situations.

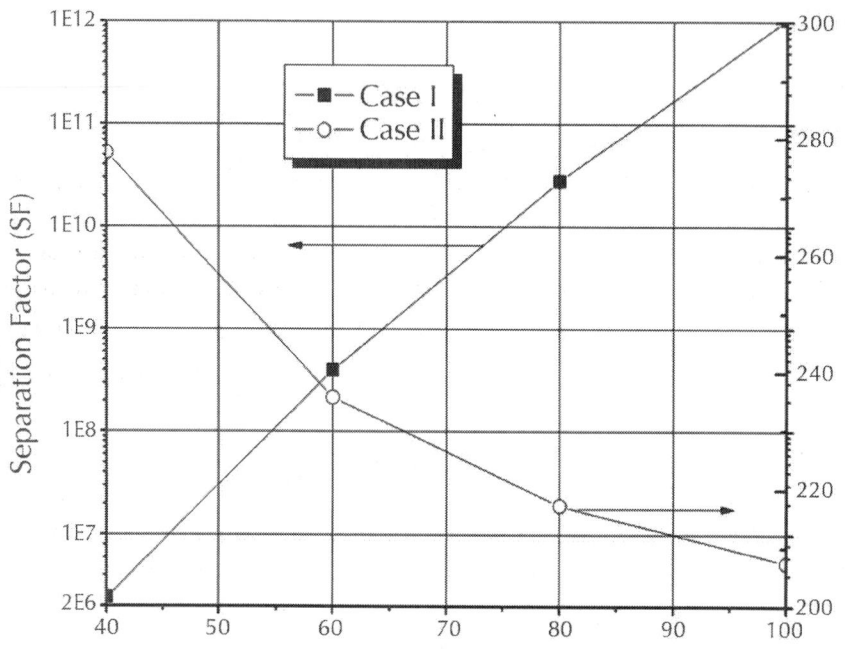

Figure 6: Global separation factor H_2O/EDC for two optimized situations.

CONCLUSIONS

Using as a case study the dehydration of 1, 2-dichloroethane (EDC) of a commercial plant that produces vinyl chloride monomer—VCM, the study aimed to evaluate the separation efficiency for two operating conditions: one with two liquid phases (Case II) and one with a single liquid phase (Case I) throughout the stages of an azeotropic distillation column.

Although very small as a percentage (less than 2.5%), the appearance of the 2nd liquid phase causes significant changes in the operation of the column and the separation achieved.

In each case, the mass transfer is at its most significant in different regions of the column: for Case I, in the upper region, for Case II, in the lower one. In fact, the transfer of mass increases when the 2nd liquid phase is not present, that is, the separation efficiency is increased when there is a single liquid phase present.

It is not a reduction in the reflux that causes the strong decrease in the Separation Factor (Case I compared to Case II); in fact, the drastic reduction in the efficiency of separation is the result of the operation with two liquid phases in some plates of the column.

ACKNOWLEDGMENTS

The authors are grateful to the Brazilian National Council for Scientific and Technological Development (CNPq) for their financial support, and also to Braskem for permission to publish the results of this study.

REFERENCES

1. G. Soave and J. A. Feliu, "Saving Energy in Distillation by Feed Splitting," Applied Thermal Engineering, Vol. 22, No. 8, 2002, pp. 889-896.
2. R. H. Perry, D. W. Green and J. O. Maloney, "Perry's Chemical Engineer's Handbook," 7th Edition, McGrawHill, New York, 1999.
3. L. Laroche, N. Bekiaris, H. W. Andersen and M. Morari, "The Curious Behaviour of Homogeneous Azeotropic Distillation—Implications for Entrainer Selection," AIChE Journal, Vol. 38, No. 9, 1992, pp. 1309-1328.
4. N. Bekiaris, E. G. Guttinger and M. Morari, "Multiple Steady States in Distillation: Effect of VL (L) E Inaccuracies," AIChE Journal, Vol. 46, No. 5, 2000, pp. 955-979.
5. T. M. Magnussen, L. Michelsen and A. A. Fredenslund, "Azeotropic Distillation Using UNIFAC," Chemical Engineering Progress Symposium Series, Vol. 56, No. 4, 1979.

6. M. Rovaglio and F. M. Doherty, "Dynamics of Heterogeneous Azeotropic Distillation Columns," AIChe Journal, Vol. 36, No. 1, 1990, pp. 39-52.
7. C. J. Wang, D. S. Wong, I.-L. Chien, R. F. Shih, S. J. Wang and C. S. Tsai, "Experimental Investigation of Multiple Steady States and Parametric Sensitivity in Azeotropic Distillation," Computer and Chemical Engineering, Vol. 21, 1997, pp. S535-S540.
8. W. L. Luyben, "Control of the Heterogeneous Azeotropic n-Butanol/Water," Energy and Fuels, Vol. 22, No. 6, 2008, pp. 4249-4258. doi:10.1021/ef8004064
9. Y. Wu and I. Chien, "Design and Control of Heterogeneous Azeotropic Column System for the Separation of Pyridine and Water," Industrial & Engineering Chemistry Research, Vol. 48, No. 23, 2009, pp. 10564-10576. doi:10.1021/ie901231s
10. M. Z. Lao and R. Taylor, "Modeling Mass-Transfer in 3- Phase Distillation," Industrial and Engineering Chemistry Research, Vol. 33, No. 11, 1994, pp. 2637-2650.doi:10.1021/ie00035a015
11. S. Widagdo and W. D. Seider, "Azeotropic Distillation," AIChE Journal, Vol. 42, No. 1, 1996, pp. 96-130.
12. A. Higler, R. Chande, R. Taylor, R. Baur and R. Krishna, "Non-Equilibrium Modeling of Three-Phase Distillation," Computers and Chemical Engineering, Vol. 28, No. 10, 2004, pp. 2021-2036. doi:10.1016/j.compchemeng.2004.04.008
13. B. P. Cairns and I. A. Furzer, "Multicomponent 3-Phase Azeotropic Distillation—Extensive Experimental Data and Simulation Results," Industrial and Engineering Chemistry Research, Vol. 29, No. 7, 1990, pp. 1349-1363. doi:10.1021/ie00103a040
14. T. L. Junqueira, M. O. S. Dias, R. Maciel Filho, M. R. W. Maciel and C. E. V. Rossel, "Simulation of the Azeotropic Distillation for Anhydrous Bioethanol Production: Study on the Formation of a Second Liquid Phase," Computer Aided

Chemical Engineering, Vol. 27, 2009, pp. 1143-1148. doi:10.1016/S1570-7946(09)70411-0

15. A. R. Silva, J. H. P. Brooman, L. R. Braga Jr., L. G. S. Vasconcelos and R. P. Brito, "Steady-State and Dynamics Behavior of an Industrial Azeotropic Distillation Column," The 6th Italian Conference on Chemical and Process Engineering, Pisa, 8-11 June 2003.

16. B. P. Guedes, M. F. Figueiredo, L. G. S. Vasconcelos, A. C. B. Araújo and R. P. Brito, "Sensitivity and Dynamic Behavior Analysis of an Industrial Azeotropic Distillation Column," Separation and Purification Technology, Vol. 56, No. 3, 2007, pp. 270-277. doi:10.1016/j.seppur.2007.02.014

17. "Azeotropic Data-III, Advances in Chemistry Series," In: R. F. Gould, Ed., Advances in Chemistry Series, Vol. 116, American Chemical Society, Washington DC, 1973, pp. 1-6.

18. Pennsylvania State University, "Dechema Chemistry Data Series," Deutsche Gesellschaft für Chemisches Aparatewesen, Frankfurt am Main, 1990.

Chapter 6

Catalyst Deactivation and Engineering Control for Steam Reforming of Higher Hydrocarbons in a Novel Membrane Reformer

Zhongxiang Chen, Yibin Yan, and
Said S.E.H. Elnashaie

Department of Chemical Engineering, Auburn University, 230 Ross Hall, Auburn, AL 36849-5127, USA

ABSTRACT

The catalyst deactivation and reformer performance in a novel circulating fluidized bed membrane reformer (CFBMR) for steam reforming of higher hydrocarbons are investigated using mathematical models. A catalyst deactivation model is developed based on a random carbon deposition mechanism over nickel reforming catalyst. The results show that the reformer has a strong tendency for carbon formation and catalyst deactivation at low steam to carbon feed ratios (<1.4 mol/mol) for high reaction temperatures (>700 K) and high pressures (>506.5 kPa). The trend is similar for the cases without and with hydrogen selective membranes. Based on this preliminary investigation, an engineering control approach, i.e., in-site control with a concept of critical/minimum steam to carbon feed ratio, is proposed and used to determine the carbon deposition free regions for both cases without and with hydrogen membranes. The comparison between the reported data and model simulation shows that the critical steam to carbon feed ratio predicted by the model agrees well with the reported industrial/experimental operating data.

INTRODUCTION

In recent years, considerable attention has been paid to the possibilities of utilizing the clean fuel hydrogen as an important energy source for the 21st century (Armor, 1999; Goltsov and Veziroglu, 2002; Ohi, 2002). In January 2003, President Bush announced in his State of the Union address $1.2 billion in research funding for developing clean, hydrogen-powered automobiles and fuel cells. Four major catalytic and non-catalytic approaches are widely used for hydrogen production, they are (1) steam reforming of hydrocarbons, (2) partial oxidation of heavy oil, (3) partial oxidation of coal and, (4) electrolysis of water (Scholz, 1993). Currently, steam reforming of hydrocarbons contributes about 50% of the world's hydrogen production (Scholz, 1993; Armor, 1999). The main advantages of the steam reforming process are: (1) it extracts the hydrogen not only from the hydrocarbons but also from the water (water resource is inexhaustible) and; (2) the reaction rate is very fast, although limited by thermodynamic equilibrium. However, this process is accompanied by unfavorable and undesired formation of different carbonaceous deposits or coke, which deactivates the catalyst, it can even destroy the reformer

(Rostrup-Nielsen 1979 and Rostrup-Nielsen 1997; Borowiecki et al., 1997; Bartholomew, 2001; Olsbye et al., 2002). By burning-off the deposited carbon with air or oxygen the catalyst is regenerated. The catalyst may be permanently deactivated by sintering (loss of surface area) during the catalyst regeneration, therefore careful control of burn-off is necessary (Trimm, 1984).

The carbon formation and catalyst deactivation during steam reforming of hydrocarbons have been intensively studied and different approaches have been developed for controlling the carbon formation. For example, it is possible to use potassium, magnesia, urania or molybdenum to improve the reforming catalysts by inhibiting the carbon formation or promoting the carbon gasification (Borowiecki et al., 1997; Trimm, 1999; Kepinski et al., 2000). Carbon formation can also be avoided by using high steam to carbon (of hydrocarbon) feed ratios (Twigg, 1989;Elnashaie and Elshishini, 1993; Christensen, 1996; Bartholomew, 2001). However, there is still no generally accepted model to describe the carbon formation and catalyst deactivation due to the complexity of the reforming process (Ren et al., 2002). Experimental data and theoretical analysis have shown that the carbon formation rate is largely dependent on the catalyst chemical composition as well as its preparation procedure (Rostrup-Nielsen, 1974; Borowiecki, 1987; Forzatti and Lietti, 1999). In 1945 Voorhies empirically described the carbon formation as a function of reaction time by the following equation (Vooehies, 1945):

$$C = k_c t^n, \qquad (1)$$

where C is the concentration of carbon formed on the catalyst, t is the reaction time, k_c is the carbon formation rate constant and n is an exponent (usually <1). Rostrup-Nielsen (1974) studied the carbon formation from higher hydrocarbons in a thermogravimetric system and suggested that the amount of carbon formed may be empirically expressed by the following equation:

$$C = k_c(t - t_0), \qquad (2)$$

where, t_0 is the induction time for carbon formation on the catalyst. In these equations, the amount of carbon formed on the catalyst is assumed to be independent of the partial pressure of hydrocarbons, which is a very unrealistic assumption.

In this paper a random carbon deposition mechanism is suggested for nickel reforming catalyst and then a catalyst deactivation model is developed, which is incorporated into a set of model equations to study the catalyst deactivation and engineering control for steam reforming of higher hydrocarbons in an earlier proposed circulating fluidized bed membrane reformer (CFBMR) (Chen and Elnashaie, 2002; Chen 2002 and Chen 2003a). Fig. 1 shows a complete schematic diagram of the proposed novel process. Inside the CFBMR there are a number of ceramic membrane tubes coated by thin palladium layer and/or dense perovskite oxygen selective membrane tubes. Between these membrane tubes the nickel reforming catalyst is fast fluidized and steam reforming of higher hydrocarbons takes place. The product hydrogen permeates selectively through hydrogen membranes and then it is carried away by sweep gas such as steam in the hydrogen membrane tubes. Air is fed into the oxygen selective membrane tubes where oxygen permeates into the reaction side for oxidative reforming of hydrocarbons. The deactivated catalyst is carried out of the reformer with the exit gases, regenerated in a catalyst regenerator by burning off the deposited carbon using excess air. Then the regenerated catalyst is separated from the gas stream in a gas–solid separator and finally recycled to the riser reformer. Because the effluent gases from the gas–solid separator are rich in carbon dioxide (CO_2), the main greenhouse gas causing global warning. In order to avoid/control this polluting emission and reduce the negative effect on the environment, as shown in the right of Fig. 1, a promising approach of dry reforming of methane is used in the downstream to capture the CO_2 for syngas production, which can be further converted into valuable fuel additives or chemicals such as methanol (El Solh et al., 2001; Verykios, 2003). In this paper only the left part CFBMR is investigated for the cases without and with hydrogen selective membranes.

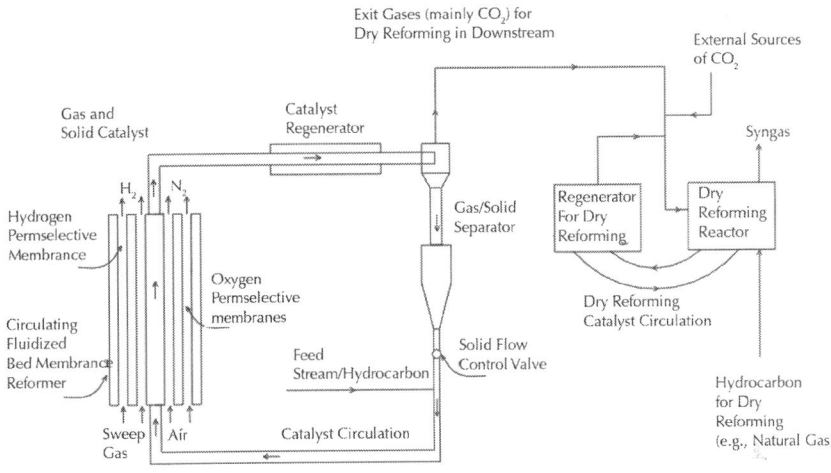

Figure 1: Schematic diagram of the novel process containing a CFBMR.

RANDOM CARBON DEPOSITION AND CATALYST DEACTIVATION MODEL

During the steam reforming of hydrocarbons on nickel reforming catalyst, three typical kinds of carbon species were identified: pyrolytic carbon, encapsulating carbon and whisker or filamentous carbon (Rostrup-Nielsen, 1979; Forzatti and Lietti, 1999; Trimm, 1999; Bartholomew, 2001). Pyrolytic carbon is usually obtained by thermal cracking of hydrocarbons above 600°C and deposition of carbon precursors. Encapsulating carbon is formed by slow polymerization of unsaturated hydrocarbons below 500°C. Whisker carbon is produced by diffusion of carbon into nickel crystals, detachment of nickel from the support and growth of whiskers with nickel on the top of the catalyst above 450°C. Both pyrolytic and encapsulating carbons cover the catalyst particle surface and therefore deactivate the catalyst. Although whisker carbon does not deactivate the catalyst directly, the accumulation of whisker carbon blocks the catalyst pores and increases the pressure drop to unacceptable levels in the reformers. Trimm (1984) suggested that the production of catalytic carbon on the nickel catalyst could best be described with the aid of Fig. 2: hydrocarbons adsorb on the catalyst surface and may react to produce gas phase products or dehydrogenated intermediates. This process continues until carbon is produced on the surface, which in turn can isomerize to other forms of carbon.

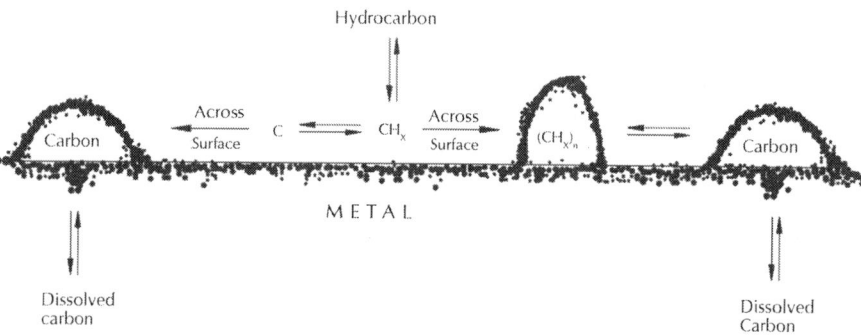

Figure 2: Mechanism of carbon formation during steam reforming of hydrocarbons (from Trimm, 1984).

Based on the above mechanism of different carbon formation over nickel reforming catalyst, a random carbon deposition and catalyst deactivation model is proposed as follows: Hydrocarbon adsorbs on the nickel reforming catalyst surface and may react to produce gas phase products or to form precursors of carbon or protocoke, which is mobile and then deposits randomly on the nickel catalyst, either on the coked or uncoked sites. The coking process can be schematically presented as follows:

$$C_nH_m \xrightarrow{r} CH_x(\text{protocoke}) \xrightarrow{r_d} \text{coke.} \tag{3}$$

Let C, C^* be the concentrations of the deposited carbon and protocoke on the catalyst; r, r_d be the protocoke formation rate and carbon deposition rate; S, S_0 be the concentrations of active sites per gram of catalyst at reaction time t and initially, respectively.

Starting with the material balance for protocoke and deposited carbon on the catalyst, we get the following two equations:

$$\frac{dC^*}{dt} = r(S, P_{C_nH_m}, \ldots) - r_d(S_0, C^*), \tag{4}$$

$$\frac{dC}{dt} = r_d(S_0, C^*). \tag{5}$$

In Eq. (4) the protocoke formation rate r involves the number of active sites at time t because the protocoke is formed from hydrocarbon, which is based on the available active sites. While in Eq. (5) the carbon deposition rate r_d involves the initial active sites S_0 and not S because of the assumption that protocoke can deposit either on the coked or uncoked sites. Considering the growth of protocoke into coke is fast and using the approximation of pseudo-steady-state for protocoke, we get

$$\frac{dC^*}{dt} = r(S, P_{C_nH_m}, \ldots) - r_d(S_0, C^*) = 0. \tag{6}$$

Since protocoke is formed from the adsorbed hydrocarbon on the catalyst, the formation rate could be presumed proportional to the concentration of adsorption sites. Then Eq. (5) may be rewritten as

$$\frac{dC}{dt} = r(S, P_{C_nH_m}, \ldots) = \frac{S}{S_0} r_0(S_0, P_{C_nH_m}, \ldots), \tag{7}$$

where r_0 is the initial protocoke formation rate. This equation proposes that the coke deposition rate in general depends on the concentration of the active sites S and S_0. If the active site coverage is the main cause of the deactivation of catalyst for steam reforming of hydrocarbons, the catalyst activity function φ(sometimes it is also called the catalyst deactivation function) may be defined by

$$\phi \equiv S/S_0. \tag{8}$$

Because the change of active sites on the catalyst equals to the loss of active sites caused by the carbon deposition, we can get

$$W_{cat} \Delta S = [-r_d(S_0, C^*)]W_{cat} \Delta t \alpha \frac{S}{S_0}, \qquad (9)$$

where W_{cat} is the catalyst weight in grams; Δt is the segment of reaction time in seconds; α is a conversion coefficient from coke concentration C to active site concentration S.

When $\Delta t \to 0$, we get the following differential equation:

$$\frac{dS}{dt} = \alpha \frac{S}{S_0}[-r_d(S_0, C^*)]. \qquad (10)$$

Substituting Eq. (5) into Eq. (10), we get

$$\frac{dS}{dt} = -\alpha \frac{S}{S_0}\frac{dC}{dt}. \qquad (11)$$

Finally, the catalyst deactivation caused by the carbon deposition may be represented by the following equation after integrating Eq. (11):

$$\phi = \exp(-\alpha_C C), \qquad (12)$$

where $\alpha_C \equiv \alpha/S_0$ is the catalyst deactivation constant. Assume that when the catalyst pores are fully filled with carbon, the catalyst will deactivate completely. For nickel catalysts, typical catalyst surface area is in the range of 20–66 m²/g-catalyst (Tottrup, 1982; Xu and Froment, 1989; Sehested et al., 2001) and pore mean radius is around 17 Å (Biswas and Do, 1987). We assume the pores are equivalent as cylinders. In the proposed CFBMR, fine catalyst particles (186μm) are used for free circulation. The reported density of industrial nickel catalyst is 2835 kg/m³ (Elnashaie and Elshishini, 1993) and the density of coke is 440:5 kg/m³ (Perry et al., 1984). Then the maximum amount of carbon that can be deposited on the catalyst is 0.16 g/g-catalyst. Ren et al. (2002) reported the experimental maximum coke content to be 16 wt% for naphtha reforming catalyst Pt–Re/Al_2O_3. Forzatti and Lietti (1999) reported that the coke deposition on the reforming catalyst may amount to 15–20% (w/w) of the catalyst. Although the specific compositions of reforming catalyst are different, the maximum carbon content estimated above is quite close to the reported data. For mathematical simplicity, suppose the catalyst will lose its 99% activity (or φ is 0.01) when the carbon content reaches 0.16g/g-catalyst. Then the catalyst deactivation constant aC is 28.8 g-catalyst/g-carbon. Therefore without carbon deposition, the catalyst does not deactivate and the catalyst activity φ is 1.0. When carbon deposition increases, the catalyst deactivates and the catalyst activity φ decreases. The larger the carbon deposition on the catalyst, the lower the catalyst activity function φ and the more significant the catalyst deactivation.

REACTIONS AND KINETICS

Many researchers used heptane as a model component for steam reforming of higher hydrocarbons (Rostrup-Nielsen, 1974; Tottrup, 1982; Christensen, 1996). In this investigation we also use heptane as a model component for higher hydrocarbons. The possible reactions and their kinetics are summarized in Table 1, which are carefully chosen from the open literatures except for the rate equation of heptane cracking. As mentioned earlier, Rostrup-Nielsen (1974) suggested a kinetic rate equation (Eq. (2)) that the amount of carbon formed on the catalyst is assumed to be independent of the partial pressure of hydrocarbons. Taking into consideration of the effect of heptane feed, we empirically correlated the Rostrup-Nielsen's experimental data and obtained the carbon formation rate equation from heptane, which is marked with a "*" in Table 1. All the reaction rates r_1 to r_9 are for reactants, as the reaction expression shown in Table 1. The carbon formation by the decomposition of carbon monoxide, i.e., the Boudouard reaction, is usually regarded as a reversible reaction. However, during the steam reforming of higher hydrocarbons, the carbon formation from hydrocarbon and methane are more important than that from carbon monoxide. Furthermore, at the presence of steam and hydrogen, carbon gasification by steam and hydrocarbon are also more important than that by CO_2. On the other hand, the kinetics of Boudouard reaction for carbon formation on the nickel catalyst reported by Tottrup (1976) is an irreversible rate equation.

In order to address this, we use another reaction rate equation r_9 shown in Table 1 for the reverse of the Boudouard reaction.

Table 1: Reactions and kinetic rate equations

Reaction	Kinetic equation	Reference
$C_7H_{16}+7H_2O \rightarrow 7CO+15H_2$	$r_1 = \dfrac{k_1 P_{C_7H_{16}}}{\left[1 + 25.2 \dfrac{P_{C_7H_{16}} P_{H_2}}{P_{H_2O}} + 0.077 \dfrac{P_{H_2O}}{P_{H_2}}\right]^2}$	Tottrup (1982)
$CO+3H_2 \rightleftharpoons CH_4+H_2O$	$r_2 = k_2 \left(\dfrac{P_{H_2}^{3}}{P_{CH_4} P_{H_2O}} - \dfrac{K_2}{P_{CO} P_{H_2}^{3}} \right) / DEN^2$	Xu and Froment (1989)
$CO+H_2O \rightleftharpoons CO_2+H_2$	$r_3 = k_3 \left(\dfrac{P_{CO} P_{H_2O}}{P_{H_2}} - \dfrac{P_{CO_2}}{K_3} \right) / DEN^2$	Xu and Froment (1989)
$CH_4+2H_2O \rightleftharpoons CO_2+4H_2$	$r_4 = k_4 \left(\dfrac{P_{CH_4} P_{H_2O}^2}{P_{H_2}^{3.5}} - \dfrac{P_{CO_2} P_{H_2}^{0.5}}{K_2 K_3} \right) / DEN^2$	Xu and Froment (1989)
$C_7H_{16} \rightarrow 7C+8H_2$	$r_5 = k_5 P_{C7H16}^{0.569}$ [a]	Rostrup-Nielsen (1974)
$CH_4 \rightleftharpoons C+2H_2$	$r_6 = \dfrac{k_6 K_{CH_4} \left(P_{CH_4} - \dfrac{P_{H_2}^2}{K_{6a}} \right)}{\left(1 + \dfrac{P_{H_2}^{1.5}}{K_{6b}} + K_{CH_4} P_{CH_4}\right)^2}$	Snoeck et al. (1997)
$2CO \rightarrow C+CO_2$	$r_7 = \dfrac{k_7 P_{CO}}{\left(1 + K_{7a} P_{CO} + K_{7b} \dfrac{P_{CO_2}}{P_{CO}}\right)^2}$	Tottrup (1976)
$C+H_2O \rightarrow CO+H_2$	$r_8 = k_8 P_{H_2O}^{0.5}$	Chen et al. (2000)
$C+CO_2 \rightarrow 2CO$	$r_9 = k_9 P_{CO_2}^{0.5}$	Chen et al. (2000)
where, $DEN = 1 + K_{CO}P_{CO} + K_{H2}P_{H2} + K_{CH4}P_{CH4} + K_{H2O}P_{H2O}/P_{H2}$		

[a] Empirically obtained from the experimental data reported by Rostrup-Nielsen (1974).

MATHEMATICAL MODELING AND SIMULATION CONDITIONS

Due to the high gas–solid velocity (~3m/s) in the circulating fluidized bed reformer, we assume that plug flow model applies in this novel CFBMR for both the gas and solid phases. Thus the CFBMR is modeled as a plug flow reactor (PFR) with co-current flow in the reactor and membrane sides. The other major assumptions for the mathematical model are as follows:

- Steady-state operation in the reaction and hydrogen membrane sides.
- The palladium based composite membranes are 100% selective for hydrogen only.
- There is no slip between the solid and gases, both are in plug flow.
- The heat capacities of the components are constant.
- The reformer and hydrogen membranes are operated at constant pressure.
- The reformer is simulated under isothermal conditions.
- The deactivated catalyst is fully regenerated before recycling to the riser reformer.
- The hydrogen selective membranes are not affected by carbon formation.

The steady-state model equations for material balance in reaction side are given by

$$\frac{dF_i}{dl} = \rho_C(1-\varepsilon)A_f \sum_{j=1}^{9} \sigma_{i,j} r_j - a J_{H_2} \pi N_{H_2} d_{H_2}, \tag{13}$$

where a is a control index, when component i is hydrogen, $a=1$, otherwise, $a=0$.

The material balance equation in hydrogen selective membrane tubes is given by

$$\frac{dF_{H_2,P}}{dl} = \pi N_{H_2} d_{H_2} J_{H_2}. \tag{14}$$

For palladium based hydrogen selective membranes, the hydrogen permeation flux can be calculated by the following equation (Shu and Kaliaguine, 1994; Barbieri and Di Maio, 1997):

$$J_{H_2} = \frac{2.003 \times 10^{-5}}{\delta_{H_2}} \exp\left(\frac{-15,700}{RT}\right)$$

$$\times \left(\sqrt{p_{H_2,r}} - \sqrt{p_{H_2,p}}\right) \frac{\text{mol}}{\text{m}^2 \text{ s}}. \tag{15}$$

Since the catalyst deactivation occurs when carbon deposits, the rate equations are reformulated accordingly by introducing the catalyst activity function to the reaction rates as follows:

$$r_j = r_{j0}\phi_j, \tag{16}$$

where ϕ_j is the specific catalyst activity function for the jth reaction, which is calculated using Eq. (12) or equal to 1.0 depending on whether the jth reaction is affected by the catalyst deactivation. In this preliminary investigation, only the last two reaction rate equations r_8 and r_9 are considered unaffected by the catalyst deactivation because the uncatalyzed carbon gasification kinetics are used in Table 1. r_{j0} is the initial reaction rate with fresh catalyst.

Using the above reaction kinetic equations, catalyst activity function and reactor model equations, the catalyst deactivation and hydrogen production in the CFBMR are investigated. Unless otherwise specified, the simulation is performed at the following standard conditions summarized in Table 2.

Table 2: Standard simulation conditions

CFBMR construction parameters and catalyst properties	
Length of the reformer and membrane tubes	2m
Diameter of the reformer	0:0978 m^a
Diameter of hydrogen selective membrane tubes	0:00498 m^b

Thickness of palladium layer on hydrogen membrane tubes	20m[c]
Catalyst particle density	2835 kg=m3[a]
Mean catalyst particle size	186m[b]
Solid fraction in circulating fluidization bed	0.2[d]
Process gas feed and reaction conditions [e]	
Heptane feed rate	0:178 mol/s
Steam feed rate	2:5 mol/s
Steam to carbon ratio	2 mol/mol
Reaction temperature	832 K
Reaction pressure	1013 kPa
Pressure in hydrogen selective membrane tubes	101 3 kPa
Sweep gas feed rate in hydrogen selective membrane tubes	0 278 mol/s

[a]Based on Elnashaie and Elshishini (1993).
[b]Based on Adris et al. (1994).
[c]Based on Shu and Kaliaguine (1994).
[d]Based on Kunii and Levenspiel 1990 and Kunii and Levenspiel 1997.
[e]Based on Tottrup (1982) and checked the CFBMR is simulated at circulating fluidization regime.

RESULTS AND DISCUSSION

Catalyst Deactivation and CFBMR Performance without Hydrogen Selective Membranes

In this paper the catalyst deactivation and CFBMR performance for the steam reforming of heptane is investigated. The simulation is performed under isothermal condition. As shown later some scales of the three-dimension plots are reversed in the direction for the better view purpose. First, the catalyst deactivation and CFBMR performance are investigated at 1013 kPa with different steam to carbon (S/C) feed ratios and reaction temperatures for the case without hydrogen selective membranes.

Fig. 3 shows the catalyst activity as a function of steam to carbon (of heptane) feed ratio and reaction temperature under the reaction pressure of 1013 kPa. The investigated range of S/C feed ratio is 0–4.4 mol/mol and the range of temperature is 623–923K. Fig. 4 shows the carbon content on the catalyst for this investigation. At low S/C feed ratios and high reaction temperatures, the nickel reforming catalyst is deactivated significantly since a lot of heptane cracks to form carbon. The catalyst activity can be as low as 0.69 shown in Fig. 3.

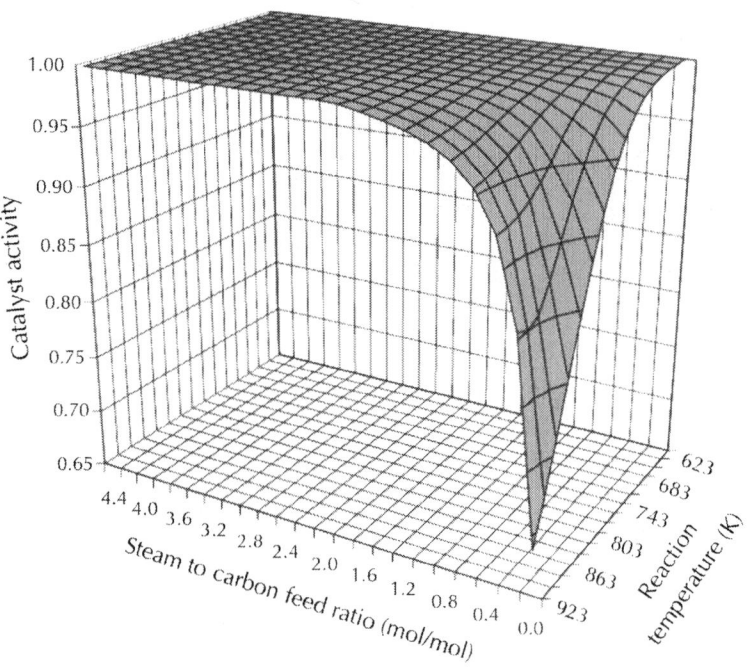

Figure 3: Catalyst activity as a function of steam to carbon feed ratio and reaction temperature for the case without hydrogen selective membranes at 1013 kPa.

The carbon content on the catalyst shown in Fig. 4 is up to 0.0130 g/g -catalyst or 1.3 wt% for the case without hydrogen selective membranes. At lower temperatures 623–683, the catalyst deactivation is negligible because of the limited carbon deposition at these lower temperatures, as shown in Fig. 4. The catalyst activity increases with the increase of S/C feed ratio and with the decrease of reaction temperature. At the corner where S/C feed ratio is less than 1.4mol/mol and the reaction temperature is higher than 700K, the reforming reactions have a strong tendency for carbon deposition on the nickel reforming catalyst and therefore causing significant catalyst deactivation shown in Fig. 3.

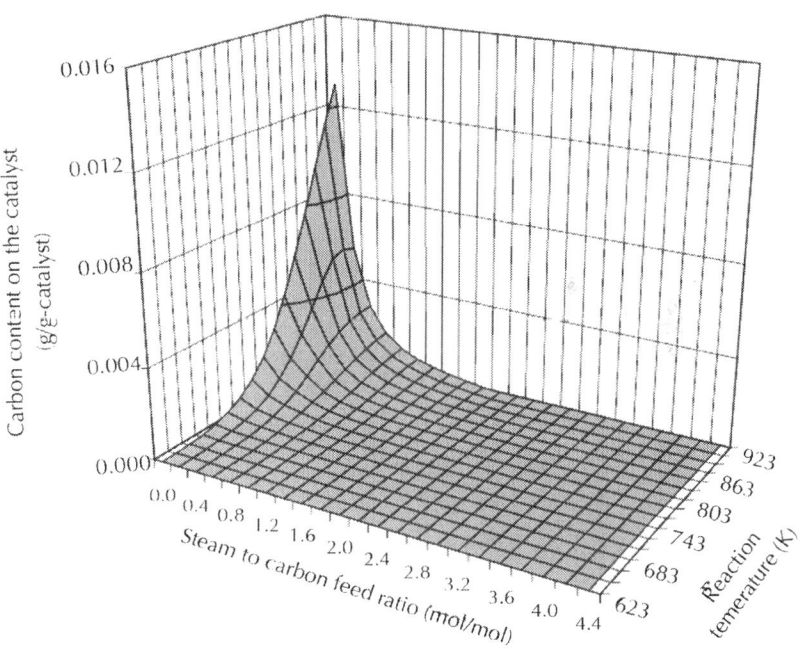

Figure 4: Carbon content on the catalyst as a function of steam to carbon feed ratio and reaction temperature for the case without hydrogen selective membranes at 1013 kPa.

Fig. 5 and Fig. 6 show the conversion of heptane (defined as the total moles of heptane converted per mol of heptane fed) and total yield of hydrogen (defined as the total moles of hydrogen produced, both in reaction side and hydrogen selective membrane tubes, per mole of heptane fed) under different S/C feed ratios and reaction temperatures. For most reaction conditions heptane is fully converted by steam reforming and heptane cracking. At lower temperatures and/or lower S/C feed ratios, for example, at 623–723K and/or at S/C feed ratio of 0–1.0 mol/mol, the conversion of heptane is relatively small. The conversion of heptane increases when reaction temperature or S/C feed ratio increases. Fig. 6 shows that the total yield of hydrogen is non-monotonic with respect to the reaction temperature when the S/C feed ratio is higher than 0.4 mol/mol. This phenomenon has been extensively investigated by Chen et al. (2003b). It is mainly caused by the strong methanation reaction around 723K in the steam reforming of heptane system. At low temperatures such as 623–723K, the steam reforming of heptane reaction dominates the system and the methanation reaction is relatively negligible. Around 723K the methanation reaction becomes more significant and consumes a lot of hydrogen produced from heptane steam reforming, causing a large decrease in the yield of hydrogen. However, at high temperatures such as 823K or higher, the steam reforming of methane, the reverse process of methanation, become more and more important, thus it decreases the formation of methane and enhances the production of hydrogen. As a result, the non-monotonic behavior in the yield of hydrogen with respect to the reaction temperature appears (Chen et al., 2003b). Fig. 7 shows the yield of methane in the reforming system. The higher the tendency for methanation, the lower the yield of hydrogen. For heptane steam reforming, the maximum theoretical yield of hydrogen is 22 according to the following complete reforming reaction in which the final products are CO_2 and hydrogen:

$$C_7H_{16} + 14H_2O \rightarrow 7CO_2 + 22H_2. \quad (17)$$

Obviously, because of the reversibility associated with the methane steam reforming reaction (or methanation) and water gas shift reaction, the production of hydrogen is usually limited by thermodynamic equilibrium, resulting in low yield of hydrogen, even at high S/C feed ratio. For example, at S/C feed ratio of 4.0 mol/mol, the total yields of hydrogen are 13.624 at 623K, 2.844 at 723K, 6.650 at 823K and 11.760 at 923K, respectively.

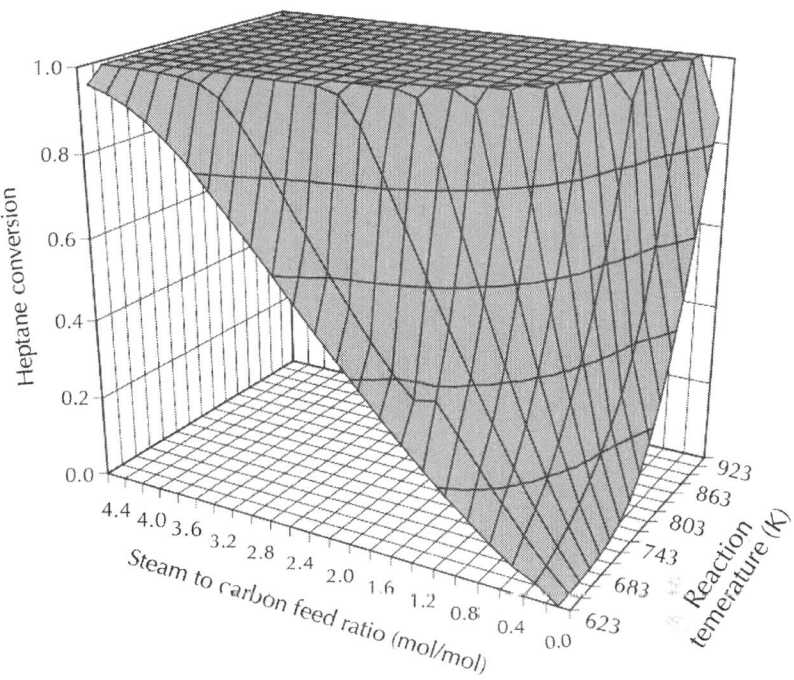

Figure 5: Heptane conversion as a function of steam to carbon feed ratio and reaction temperature for the case without hydrogen selective membranes at 1013 kPa.

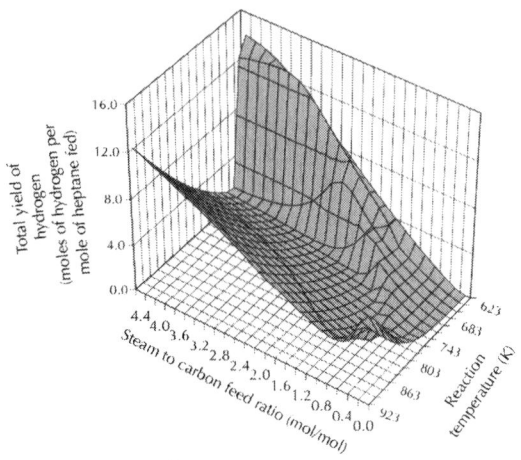

Figure 6: Total yield of hydrogen as a function of steam to carbon feed ratio and reaction temperature for the case without hydrogen selective membranes at 1013 kPa.

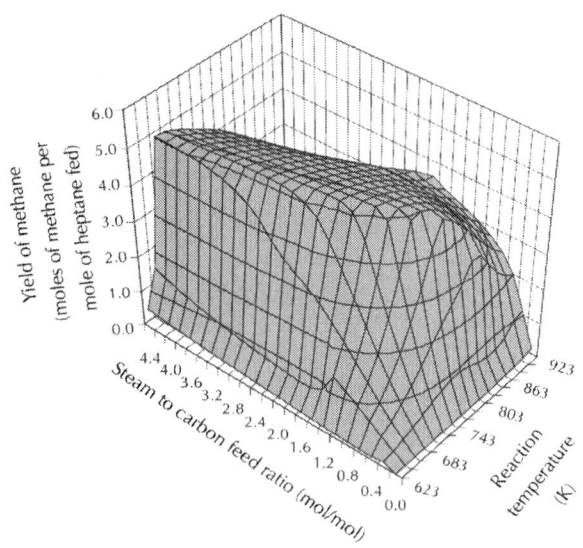

Figure 7: Yield of methane as a function of steam to carbon feed ratio and reaction temperature for the case without hydrogen selective membranes at 1013 kPa.

Next, the catalyst deactivation and CFBMR performance are investigated at 823K with different S/C feed ratios and reaction pressures for the case without hydrogen selective membranes. The investigated range of S/C feed ratio is the same, 0–4.4 mol/mol. The range of reaction pressure is 101.3–3140.3 kPa. Fig. 8 shows the catalyst activity as a function of S/C feed ratio and reaction pressure at 823K. For most of the operation conditions, the catalyst activity is high (close to 1.0), which means low carbon content on the reforming catalyst and insignificant catalyst deactivation. At S/C feed ratio of 1.4 mol/mol or higher, the catalyst activity is higher than 0.972. Accordingly, the carbon content on the catalyst is below 0.001 g/g-catalyst or 0.1 wt%. At a corner where S/C feed ratio is less than 1.4 mol/mol and reaction pressure is higher than 506.5 kPa, the reforming reactions have a strong tendency for carbon formation and deposition on the nickel reforming catalyst, causing a significant catalyst deactivation. The carbon content can reach as high as 0.0181 g/g -catalyst at 3140 3 kPa for the special condition without steam (i.e., S/C feed ratio is 0 mol/mol), resulting in a very low catalyst activity, about 0.594. The catalyst activity increases when the S/C feed ratio increases or when the reaction pressure decreases. At high S/C feed ratio, the steam reforming of hydrocarbons (heptane and by-product methane) dominates the reforming system and the high excess steam feed enhances the carbon gasification, which suppresses the carbon deposition on the nickel reforming catalyst and therefore decreases the catalyst deactivation.

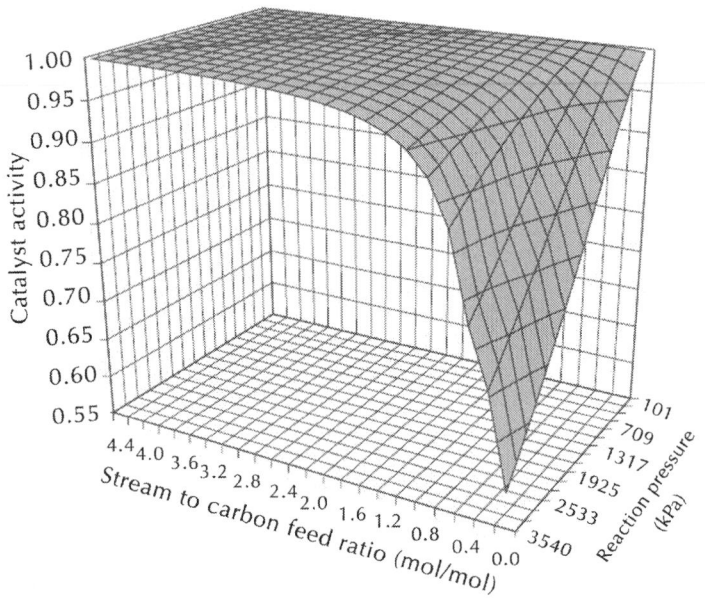

Figure 8: Catalyst activity as a function of steam to carbon feed ratio and reaction pressure for the case without hydrogen selective membranes at 823 K.

Fig. 9 shows the total yield of hydrogen as a function of S/C feed ratio and reaction pressure at 823 K for the case without hydrogen membranes. At very low S/C feed ratio, for example, 0–0.4 mol/mol, the yield of hydrogen increases when the reaction pressure increases, while above 0.4 mol/mol, the yield of hydrogen increases with the increase of the S/C feed ratio and decreases with the increase of the reaction pressure. This can be explained as follows: at low S/C feed ratio 0–0.4 mol/mol, the heptane cracking reaction is dominating in the system. Because this cracking reaction is irreversible, the high pressure will not limit the conversion of heptane for cracking. Since the reaction order of the cracking of heptane with respect to the partial pressure of heptane is 0.569 shown in Table 1, high operating pressure gives high cracking rate of heptane. As a result, the yield of hydrogen increases when the reaction pressure increases. However, when

S/C feed ratio increases, the steam reforming of heptane becomes important and also the methanation and water gas shift reactions. The methanation or steam reforming of methane and water gas shift reaction are fast reversible reactions, which are strongly affected by the thermodynamic equilibrium. Since steam reforming of heptane is accompanied with an increase in molecule number, the higher the operating pressure, the higher the effect of the thermodynamic equilibrium. Thus the yield of hydrogen decreases when the reaction pressure increases. This phenomenon also indicates that the thermodynamic equilibrium limits the production of hydrogen in steam reforming system.

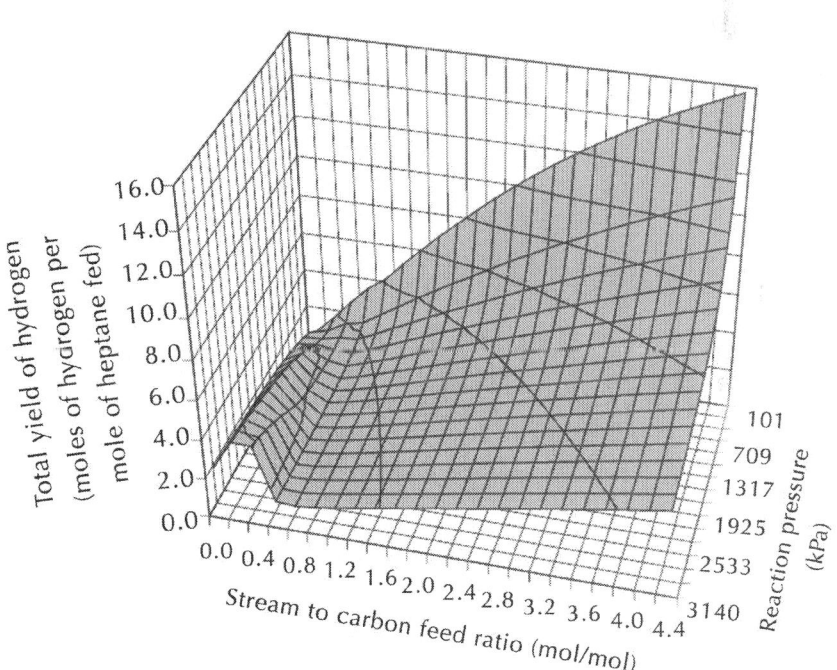

Figure 9: Yield of hydrogen as a function of steam to carbon feed ratio and reaction pressure for the case without hydrogen selective membranes at 823K.

Catalyst Deactivation and CFBMR Performance with Hydrogen Selective Membranes

In this section 20 hydrogen selective membranes are used to investigate their effect on the catalyst deactivation and CFBMR performance. Fig. 10 shows the catalyst activity for this case. The trend is similar to the earlier case without hydrogen selective membranes. The catalyst is also deactivated significantly at low S/C feed ratio and high reaction temperatures. At most reforming conditions the effect of carbon deposition on the catalyst deactivation is negligible. While at the corner where S/C feed ratio is less than 1.6 mol/mol and the reaction temperature is higher than 700K, the catalyst activity decreases significantly with the decrease of the S/C feed ratio and with the increase of the reaction temperature. The difference between both cases without and with hydrogen selective membranes is only the magnitude of the catalyst activity. With 20 hydrogen membranes the lowest catalyst activity is 0.59. Accordingly, the maximum carbon content on the catalyst is 0.0183 g/g -catalyst or 1.8 wt %. While in the earlier case without hydrogen selective membranes the lowest catalyst activity function is 0.69 shown in Fig. 3 or the maximum carbon content is 0.0130 g/g -catalyst (1.3 wt%) shown in Fig. 4. The carbon content for the case with 20 hydrogen selective membranes increases by 40.8%. The result can be explained by the effect of hydrogen selective membranes on the reactions, especially on the carbon formation and carbon gasification in the CFBMR. Table 3 summarized the possible effects of hydrogen selective membranes on the directions of different reactions. For those irreversible reactions, the reaction direction will not be affected by the use of hydrogen selective membranes. While for the reversible reactions, the use of hydrogen membranes will be favorable or "shift" some reactions to certain directions as clearly shown inTable 3. Although the use of the hydrogen selective membranes will be favorable for the steam reforming of methane (by-product of steam reforming of heptane via methanation) or suppress the formation of methane, the strong methanation can still produce much methane for carbon formation, leading to a little higher carbon content on the nickel reforming catalyst. At the extreme condition without steam, i.e., S/C feed ratio of 0 mol/mol, the carbon produced by heptane cracking will not be gasified by steam. However, the product hydrogen can react with carbon to form methane at high temperatures.

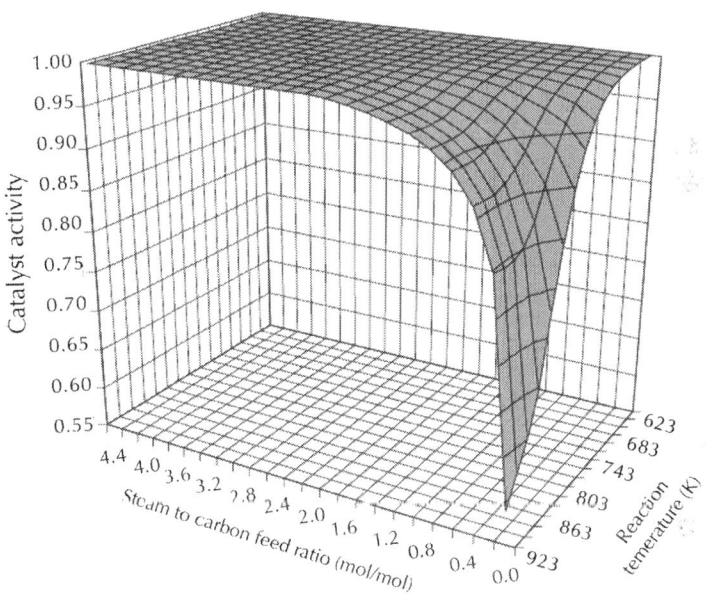

Figure 10: Catalyst activity as a function of steam to carbon feed ratio and reaction temperature for the case with 20 hydrogen selective membranes at 1013 kPa.

he use of hydrogen selective membranes decreases the concentration of hydrogen in the system, as a result the carbon gasification by hydrogen is lower and the carbon content is higher for the case with hydrogen selective membranes. Therefore, the catalyst activity is lower than the case without hydrogen selective membranes, which is shown in Fig. 10. As mentioned above, the use of hydrogen selective membranes "shifts" the reversible reactions to the direction for hydrogen production. Therefore, the yield of hydrogen shown in Fig. 11 is significantly improved using hydrogen selective membranes. For example, at S/C feed ratio of 2 mol/mol with 823K and 1013 kPa, the yield of hydrogen is 3.726 for the case without hydrogen selective membranes and 19.298 for the case with hydrogen selective membranes, which is close to the theoretical maximum yield of hydrogen 22 shown by Eq. (17). The improvement is about 418% due to the "break" of the thermodynamic equilibrium limitation.

Table 3: Effect of hydrogen selective membranes on the reaction directions

Reactions	Effect of hydrogen membranes on the reaction direction
$C_7H_{16}+7H_2O \rightarrow 7CO+15H_2$	NA[a]
$CO+3H_2 \rightleftharpoons CH_4+H_2O$	$\leftarrow$[b]
$CO+H_2O \rightleftharpoons CO_2+H_2$	$\rightarrow$
$CH_4+2H_2O \rightleftharpoons CO_2+4H_2$	$\rightarrow$
$C_7H_{16} \rightarrow 7C+8H_2$	NA
$CH_4 \rightleftharpoons C+2H_2$	$\rightarrow$
$2CO \rightarrow C+CO_2$	NA
$C+H_2O \rightarrow CO+H_2$	NA
$C+CO_2 \rightarrow 2CO$	NA

[a]NA=not affected.
[b]Left or right arrows mean "favorable" for this reaction direction.

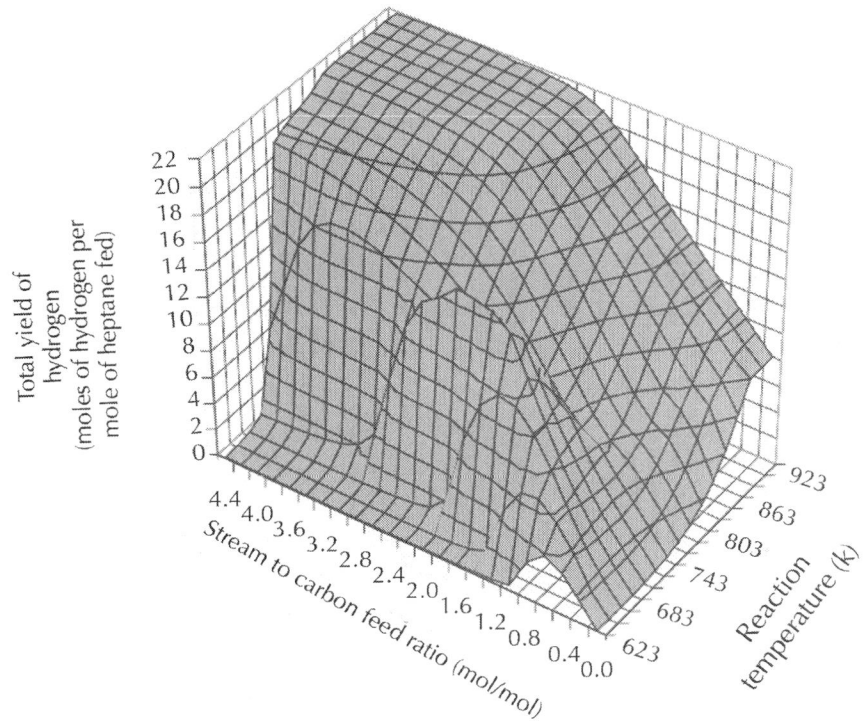

Figure 11: Yield of hydrogen as a function of steam to carbon feed ratio and reaction temperature for the case with 20 hydrogen selective membranes at 1013 kPa.

Fig. 12 shows the catalyst activity at 823K as the function of S/C feed ratio and reaction pressure, respectively. The catalyst activity profile is similar to that shown in Fig. 8. While the lowest catalyst activity is 0.467 at S/C feed ratio of 0 mol/mol and 3140 kPa. The carbon content at this condition is 0.0264 g/g -catalyst for the case with hydrogen selective membranes. The carbon content is 45.9% higher than the case without hydrogen selective membranes at the same operation condition. Fig. 13 shows the total yield of hydrogen as a function of S/C feed ratio and reaction pressure for the case with hydrogen selective membranes. Compared to the hydrogen yield profile shown in Fig. 9, obviously, the thermodynamic equilibrium limitation for hydrogen production due to the high operating pressure is eliminated with hydrogen selective membranes. For example, at S/C feed ratio of 2 mol/mol and 3140 kPa, the total yield of hydrogen is 2.317 for the case without hydrogen selective membranes, while with hydrogen selective membranes, the total yield of hydrogen is 20.687. In Fig. 9 the yield of hydrogen decreases when operating pressure increases due to the fact that steam reforming of heptane is accompanied with an increase in molecule number. However, using hydrogen selective membranes, the yield of hydrogen increases when operating pressure increases. The improvement using hydrogen selective membranes at high reaction pressure is significant. Fig. 13 also shows that the yield of hydrogen increases when the S/C feed ratio increases. However, at high S/C feed ratios, for example, 3.0 mol/mol or higher, the difference in hydrogen yield with different operating pressures is rather small. The total yield of hydrogen shown in Fig. 13 is very close to the theoretical maximum yield of hydrogen of 22 at the region where S/C feed ratio is higher than 3 mol/mol and reaction pressure is higher than 506.5kPa.

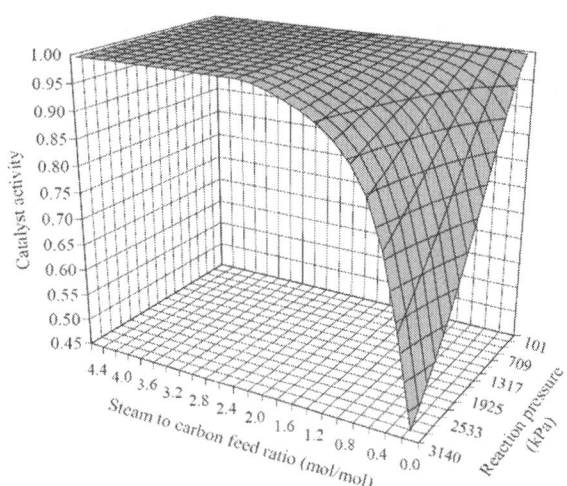

Figure 12: Catalyst activity as a function of steam to carbon feed ratio and reaction pressure for the case with 20 hydrogen selective membranes at 823K.

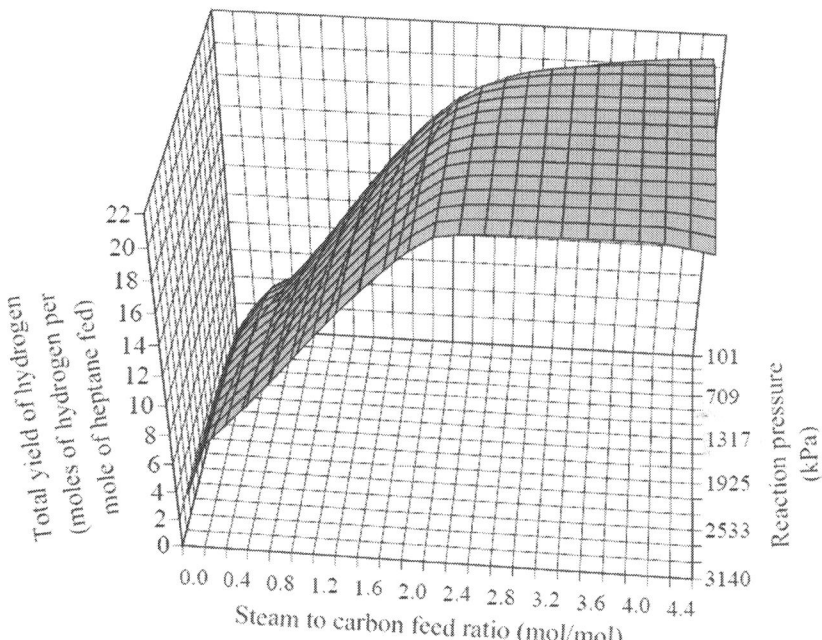

Figure 13: Total yield of hydrogen as a function of steam to carbon feed ratio and reaction pressure for the case with 20 hydrogen selective membranes at 823K.

Engineering Control for Carbon Deposition and Catalyst Deactivation

As mentioned earlier, there are several approaches to control the carbon formation on the steam reforming catalyst. These approaches may be classified into three groups according to their control stages. The first one is the well-designed optimal catalyst containing small amount of dopants such as Pt, Ir, Sn, Pb, Ge, As, Bi, Mo, Ag, etc. (Trimm, 1999). We may call it pre-reforming control or catalyst control. It usually takes a relatively long time to obtain such optimal catalysts. The second one is the in-site carbon formation suppression or gasification by steam, hydrogen, oxygen, etc. We may call it in-site control. The third one is the deactivated catalyst regeneration such as burn-off using oxygen or air in the regenerator. We may call it post-reforming control. All of them are widely used in the catalyst design, production, utilization and regeneration. In this section we focus on the second approach, the in-site control. As shown earlier, the carbon content on the catalyst can be well-controlled without significant catalyst deactivation at certain high S/C feed ratios. In order to provide a practical carbon free reforming condition for the novel CFBMR, we investigated the catalyst activity as a function of reaction temperature, pressure and S/C feed ratio. The carbon deposition free boundary is defined as the critical/minimum S/C feed ratio that makes the carbon content on the catalyst practically negligible (close to zero), which means that the catalyst activity is very high (close to 1.0). Through preliminary investigation, the critical catalyst activity for this kind of carbon deposition free boundary simulation is chosen 0.995. Then by incrementing the S/C feed ratio in small steps at a given reaction temperature and pressure (other reforming conditions are the same as listed in Table 2), the point at which the catalyst activity is equal to critical value of 0.995 can be precisely determined. Thus it is possible to describe the carbon deposition free boundary for heptane steam reforming in CFBMR by determining a series of critical S/C feed ratios as functions of reaction temperature and reaction pressure.

Fig. 14 and Fig. 15 show the critical S/C feed ratios for the cases without and with 20 hydrogen selective membranes. In these two figures, the region above the surface can be considered as the carbon deposition free zone. While below this surface it is considered as the carbon deposition zone in which the catalyst activity is smaller than 0.995 and the catalyst deactivates significantly. Thus it is possible to use these findings to guide the practical operations for the novel CFBMR, especially with regard to the carbon formation and catalyst deactivation. Generally, the critical S/C feed ratio increases with the increases of the reaction temperature and reaction pressure. Fig. 16 shows the difference of critical S/C feed ratios between these two cases with and without

20 hydrogen selective membranes. At lower reaction temperatures 623–723K and lower pressures 101–1317 kPa, the critical S/C feed ratios for the case with hydrogen selective membranes are higher than the case without hydrogen selective membranes. While at the other conditions where reaction temperature is higher than 723K and pressure is higher than 1317kPa, the critical S/C feed ratios for the case with hydrogen selective membranes are smaller than the case without hydrogen selective membranes. This interesting phenomenon can be explained by the effect of the removal of product hydrogen on the reversible steam reforming system. Chen and his coworkers have shown theoretically that the steam reforming rate of heptane is non-monotonic with respect to the partial pressure of hydrogen (Chen et al., 2003a). That is, the steam reforming rate of heptane increases when the partial pressure of hydrogen increases from 0 to 25.3kPa and then decreases after 25.3kPa. At the entrance of the reformer, the partial pressure of hydrogen is usually smaller than 25.3kPa. But the steam reforming of heptane is a fast reaction at high temperature and pressure, which supplies a lot of hydrogen near the entrance of the reformer. As a result the partial pressure of hydrogen increases quickly. Although the removal of hydrogen decreases the partial pressure of hydrogen in the reaction side, the partial pressure of hydrogen is still higher than 25.3kPa due to the continuous fast production of hydrogen at high reaction temperatures and pressures. The removal of hydrogen increases the steam reforming rate of heptane and decreases the partial pressure of heptane (or the carbon formation rate from heptane decreases) at high reaction temperatures and pressures, leading to a higher reaction rate ratio between steam reforming of heptane and carbon formation from heptane cracking. Therefore, the amount of carbon formed from heptane for the case with hydrogen selective membranes is smaller than the case without hydrogen membranes at high reaction temperatures (>723K) and pressures (>1317kPa). Secondly, the removal of hydrogen shifts the reversible steam reforming of methane and water gas shift reaction to the direction for hydrogen production. Because methane and carbon monoxide are the alternative carbon formation sources in the heptane steam reforming system, the shift of these reversible

reactions to hydrogen production also makes the concentrations of methane and carbon monoxide much lower than the case without hydrogen selective membranes. As a result it suppresses the carbon formation from these two reforming by-products methane and carbon monoxide. Although it may also shifts the carbon formation by methane cracking at certain extent, the amount of carbon formed from methane is relatively smaller than the case without hydrogen selective membranes. Then the necessary amount of steam for the carbon complete gasification is smaller. Thus the critical S/C feed ratio for the case with hydrogen selective membranes is smaller than the case without hydrogen selective membranes at high reaction temperatures (>723K) and high pressures (>1317kPa), which is shown in Fig. 16.

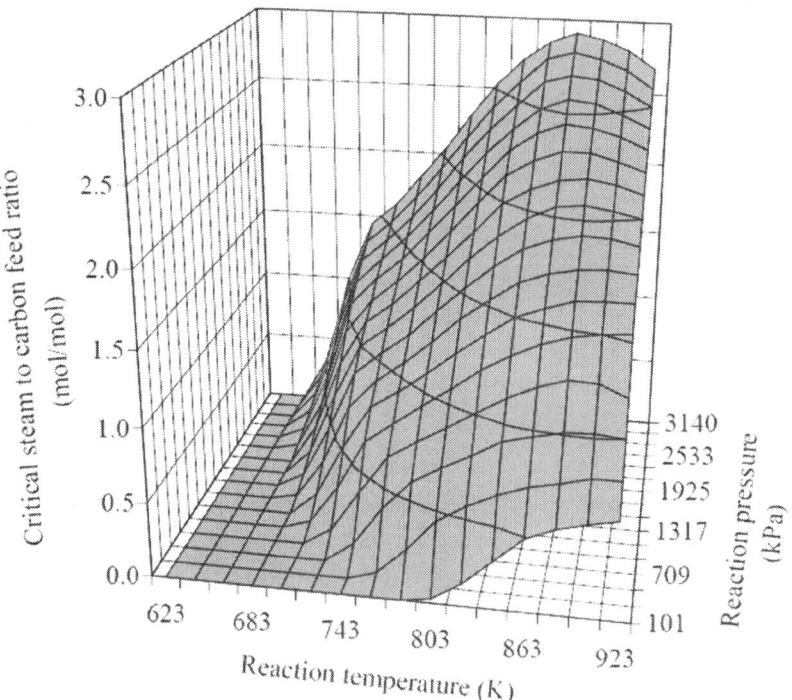

Figure 14: Carbon deposition free boundary for the case without hydrogen membranes.

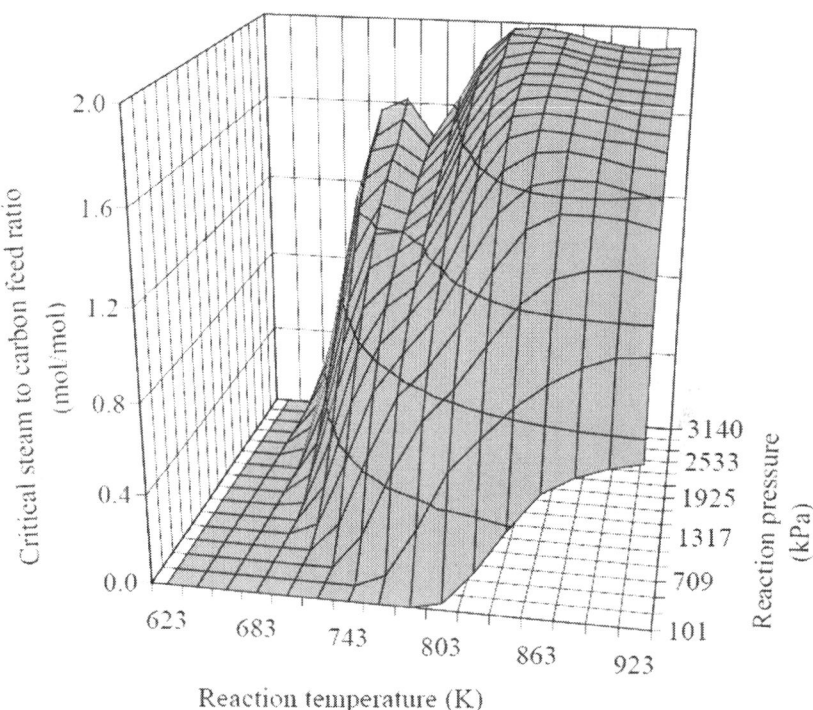

Figure 15: Carbon deposition free boundary for the case with 20 hydrogen membranes.

As shown above it is possible to use these findings in the present investigation to guide the practical operation for the novel CFBMR regarding the carbon formation and catalyst deactivation. Although the reformer configuration is quite different from the other previous steam reformers, it is still possible to use the reported industrial/experimental data to check these findings. Table 4 shows some examples of comparison between the reported data and the model simulation. The critical S/C feed ratios in the last column of Table 4 are the model simulation results. They are obtained at the conditions with same reaction temperature and pressure for the reported industrial/experimental data. The reaction temperature is usually the average temperature from these reported data since there are temperature profiles in industrial or experimental reactors for the highly endothermic steam reforming. If no such information reported, the high temperature is usually used.

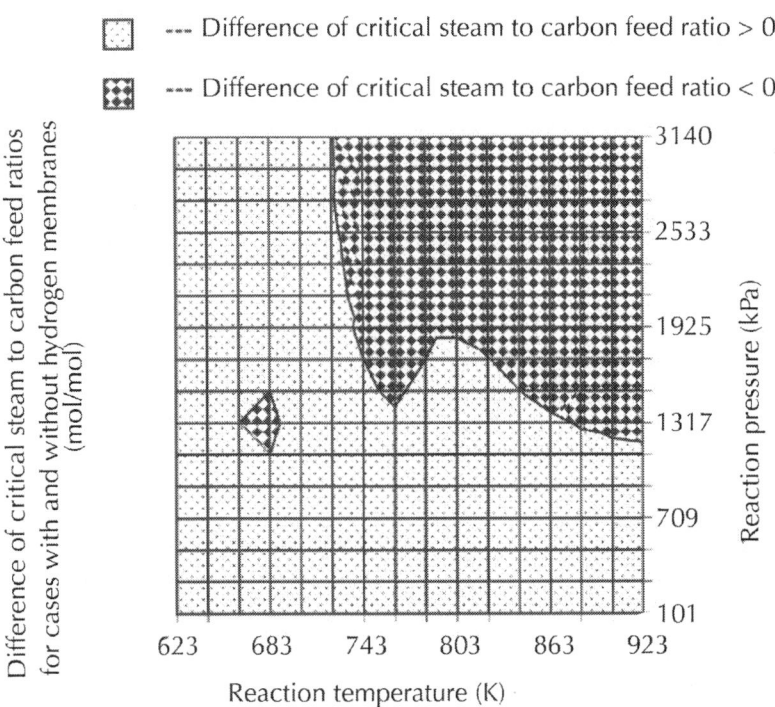

Figure 16: Difference of critical steam to carbon feed ratio for the cases with and without hydrogen selective membranes (difference of critical steam to carbon feed ratio=critical steam to carbon feed ratio for the case with hydrogen membranes−critical steam to carbon feed ratio for the case without hydrogen membranes).

The lowest operating S/C feed ratios from the reported literatures can be regarded as the values close to the minimum S/C feed ratios. From the comparison of the last two columns in Table 4, it seems that the critical S/C feed ratio predicted from the model simulation agrees well with the reported industrial/experimental data. For example, Phillips 1969 and Phillips 1970 reported their experimental S/C feed ratio is about 1.429mol/mol at 14.7atm and773K, which is the necessary ratio for satisfactory continuous operation of the industrial process. This necessary S/C feed ratio of 1.429mol/mol is very close to the model predicted critical S/C feed ratio of 1.251mol/mol.

Table 4: Examples of comparison between the reported industrial/experimental data and the model simulation

Reference	Hydrocarbon feed	Temperature (K)	Pressure (atm)	S/C feed ratio (mol/mol)	Critical S/C feed ratio by model simulation
Phillips 1969 and Phillips 1970	n-Heptane	773	14.7	1.429	1.251
Rostrup-Nielsen (1977)	Naphtha	~800	24.9	2.4–3.0	2.17
Tottrup (1982)	n-Heptane	773	20	1.9–7.4	1.57
Christensen (1996)	Naphtha	773	26	2.5–4.0	1.91

CONCLUSIONS

The nickel catalyst deactivation and CFBMR performance during steam reforming of heptane are investigated using an overall mathematical model including a random carbon deposition and catalyst deactivation model. Palladium based hydrogen selective membranes are used for the removal of product hydrogen, which "breaks" the thermodynamic equilibrium limitations associated with the reversible reforming reactions. As a result the yield of hydrogen with hydrogen selective membranes is much higher than the case without hydrogen selective membranes. The simulations show that the steam reforming of heptane has a strong tendency for carbon formation and deposition at low steam to carbon feed ratios (<1.4mol/mol) for high reaction temperatures (>700K) and pressures (>506.5kPa), which tends to deactivate the nickel reforming catalyst significantly. The effects of hydrogen selective membranes on the carbon deposition and catalyst deactivation are also investigated and the results are similar for both cases without and with hydrogen selective membranes. The catalyst activity decreases when steam to carbon feed ratio decreases, reaction temperature increases or reaction pressure increases. An engineering control approach, in-site control with a concept of critical/minimum steam to carbon feed ratio is suggested and used to determine the carbon deposition free boundary for both cases without and with hydrogen selective membranes in the CFBMR. It is found that at low reaction temperatures 623–723K and pressures 101–1317kPa, the critical steam to carbon feed ratios for the case with hydrogen selective membranes are higher than the case without hydrogen selective membranes. While at the other conditions where reaction temperature is higher than 723K and pressure is higher than 1317kPa, the critical steam to carbon feed ratios for the case with hydrogen selective membranes are smaller than the case without hydrogen selective membranes. The comparison between the reported data and model simulation shows that the critical S/C feed ratio predicted from the model agrees well with the reported industrial/experimental operating data. Thus it is possible to use these findings in the present investigation to guide the practical operation for the novel CFBMR regarding the carbon formation and catalyst deactivation as well as for other steam reformers.

ACKNOWLEDGMENTS

This work was financially supported by Auburn University, Grant Number 2-12085.

REFERENCES

1. Adris, A.M., Lim, C.J., Grace, J.R., 1994. The 2uidized bed membrane reactor (FBMR) system: a pilot scale experimental study. Chemical Engineering Science 49, 5833–5843.
2. Armor, J.N., 1999. Review: The multiple roles for catalysis

in the production of H2. Applied Catalysis A: General 176, 159–176.
3. Barbieri, G., Di Maio, F.P., 1997. Simulation of methane steam reforming process in a catalytic Pd-membrane reactor. Industrial and Engineering Chemistry Research 36, 2121–2127.
4. Bartholomew, C.H., 2001. Mechanisms of catalyst deactivation. Applied catalysis A: General 212, 17–60.
5. Biswas, J., Do, D.D., 1987. A uniKed theory of coking deactivation in a catalyst pellet. Chemical Engineering Journal 36, 175–191.
6. Borowiecki, T., 1987. Nickel catalysts for steam reforming of hydrocarbons: direct and indirect factors alecting the coking rate. Applied Catalysis 31, 207–220.
7. Borowiecki, T., StasiYnska, B., Go lebiowski, A., 1997. Elects of small MoO3 additions on the properties of nickel catalysts for the steam reforming of hydrocarbons. Applied Catalysis A: General 141–156.
8. Chen, Z., Elnashaie, S.S.E.H., 2002. EScientproduct ion of hydrogen from higher hydrocarbons using novel membrane reformer. 14th World Hydrogen Energy Conference, Montreal, Canada, June 9–13.
9. Chen, C.X., Masayuki Horio, Toshinori Kojima, 2000. Numerical simulation of entrained 2ow coal gasiKers. Part I: modeling of coal gasiKcation in an entrained 2ow gasiKer. Chemical Engineering Science 55, 3861–3874.
10. Chen, Z., Prasad, P., Elnashaie, S.S.E.H., 2002. The coupling of catalytic steam reforming and oxidative reforming of methane to produce pure hydrogen in a novel circulating fast 2uidized bed membrane reformer. ACS Meeting, Orlando, FL, Fuel Chemistry Division Preprints 47 (1), 111–113.
11. Chen, Z., Yan, Y., Elnashaie, S.S.E.H., 2003a. Modeling and optimization of a novel membrane reformer for higher hydrocarbons. A.I.Ch.E. Journal 49 (5), 1250–1265.
12. Chen, Z., Yan, Y., Elnashaie, S.S.E.II., 2003b. Non-monotonic

behavior of hydrogen production from higher hydrocarbon steam reforming in a circulating fast 2uidized bed membrane reformer. Industrial and Engineering Chemistry Research 42 (25), 6549–6558.
13. Christensen, T.S., 1996. Adiabatic prereforming of hydrocarbons—an important step in syngas production. Applied Catalysis A: General 138, 285–309.
14. Elnashaie, S.S.E.H., Elshishini, S.S., 1993. Modelling, Simulation and Optimization of Industrial Fixed Bed Catalytic Reactors. Gordon and Breach Science Publishers, London, UK.
15. El Solh, T., Jarosch, K., de Lasa, H.I., 2001. Fluidized catalyst for methane reforming. Applied Catalysis A: General 210 (1–2), 315–324.
16. Forzatti, P., Lietti, L., 1999. Catalyst deactivation. Catalysis Today 52, 165–181.
17. Goltsov Victor, A., Nejat Veziroglu, T., 2002. A step on the road to hydrogen civilization. International Journal of Hydrogen Energy 27 (7–8), 719–723.
18. Kepinski, L., Stasinska, B., Borowiecki, T., 2000. Carbon deposition on Ni=Al2O3 catalysts doped with small amounts of Molybdenum. Carbon 38, 184–185.
19. Kunii, D., Levenspiel, O., 1990. Entrainment of solids from 2uidized beds: I. Hold-up of solids in the freeboard, II. Operation of fast 2uidized beds. Powder Technology 61, 193–206.
20. Kunii, D., Levenspiel, O., 1997. Circulating 2uidized-bed reactors. Chemical Engineering Science 15, 2471–2484.
21. Ohi, J., 2002. Hydrogen energy futures: scenario planning by the U.S. DOE hydrogen technical advisory panel. 14th World Hydrogen Energy Conference, Montreal, Canada, June 9–13.
22. Olsbye, U., Moen, O., Slagtern, AU ., Dahl, I.M., 2002. An investigation of the coking properties of Kxed and 2uid bed reactors during methane-to-synthesis gas reactions. Applied Catalysis A: General 228, 289–303.

23. Perry, R.H., Chilton, C.H., Kirkpatrick, S.D., 1984. Chemical Engineers' Handbook, 6th Edition. Mcgraw-Hill Book Co., New York.
24. Phillips, T.R., Mulhall, J., Turner, G.F., 1969. The kinetics and mechanism of the reaction between steam and hydrocarbons over Nickel catalysts in the temperature range 350–500°C, PartI. Journal of Catalysis 15, 233.
25. Phillips, T.R., Mulhall, J., Turner, G.F., 1970. The kinetics and mechanism of the reaction between steam and hydrocarbons over Nickel catalysts in the temperature range 350–500°C, PartII. Journal of Catalysis 17, 28.
26. Ren, X.-H., Bertmer, M., Stapf, S., Demco, D.E., Bl\umich, B., Kern, C., Jess, A., 2002. Deactivation and regeneration of a naphtha reforming catalyst. Applied Catalysis A: General 39–52.
27. Rostrup-Nielsen, J.R., 1974. Coking on Nickel catalysts for steam reforming of hydrocarbons. Journal of Catalysis 33, 184–201.
28. Rostrup-Nielsen, J., 1977. Hydrogen via steam reforming of Naphtha. Chemical Engineering Progress 9, 87.
29. Rostrup-Nielsen, J.R., 1979. Symposium on the science of catalysis and its application in industry, Sindri, India, 22–24.
30. Rostrup-Nielsen, J.R., 1997. Industrial relevance of coking. Catalysis Today 37, 225–232.
31. Scholz, W.H., 1993. Processes for industrial production of hydrogen and associated environmental elects. Gas Separation and PuriKcation 7, 131–139.
32. Sehested, J., Carlsson, A., Janssens, T.V.W., Hansen, P.L., Datye, A.K., 2001. Sintering of Nickel steam-reforming catalysts on MgAl2O4 spinel supports. Journal of Catalysis 197, 200–209.
33. Shu, B.P.A.G., Kaliaguine, S., 1994. Methane steam reforming in asymmetric Pd- and Pd-Ag/porous SS membrane reactor. Applied Catalysis A 119, 305–325.

34. Snoeck, J.W., Froment, G.F., Fowles, M., 1997. Kinetic study of the carbon Klament formation by methane cracking on a Nickel catalyst. Journal of Catalysis 169, 250–262.
35. Tottrup, P.B., 1976. Kinetics of decomposition of carbon monoxide on a supported Nickel catalyst. Journal of Catalysis 42, 29–36.
36. Tottrup, P.B., 1982. Evaluation of intrinsic steam reforming kinetic parameters from rate measurements on full particle size. Applied Catalysis 4, 377–389.
37. Trimm, D.L., 1984. Control of coking. Chemical Engineering Process 18, 137–148.
38. Trimm, D.L., 1999. Catalysts for the control of coking during steam reforming. Catalysis Today 49, 3–10.
39. Twigg, M.V., 1989. Catalyst Handbook, 2nd Edition, Wolfe Publishing Ltd, England, pp. 225–282.
40. Verykios, X.E., 2003. Catalytic dry reforming of natural gas for the production of chemicals and hydrogen. International Journal of Hydrogen Energy 28 (10), 1045–1063.
41. Vooehies Jr., A., 1945. Industrial and Engineering Chemistry 37, 318.
42. Xu, J., Froment, G.F., 1989. Froment, Methane steam reforming, methanation and water–gas shift: I. Intrinsic kinetics. Journal of AIChE 35 (1), 88–96.

Chapter 7

In Silico Bioremediation of Polycyclic Aromatic Hydrocarbon: A Frontier in Environmental Chemistry

Vito Librando[a,b] and Matteo Pappalardo[a,b]

[a]Dipartimento di Scienze Chimiche, Università di Catania, Viale A.Doria 6, 95125 Catania, Italy

[b]Research Center for Analysis, Monitoring and Minimization Methods of Environmental Risk, Chemical Science Building, Viale A.Doria 6, 95125 Catania, Italy

ABSTRACT

In recent years, the number of studies in the field of bioremediation has been growing steadily. Although a large number of studies provide information that is highly detailed and offer great amounts of knowledge on a given subject, the downside is that the hunt for more information requires the combined efforts of researchers from many areas, which are becoming increasingly difficult to attain. In this review, we present an overview of recent work investigating enzyme degradation of polycyclic aromatic hydrocarbons. In the first part, this review examines several of the new enzymes able to degrade pollutants, with special attention being given to those with a well-resolved structure. The second part explores some of the most recent work in which computational approaches, such as molecular dynamics, docking, density functional theory and database retrieval, have been employed to study enzymes with specific bioremediation activities.

GRAPHICAL ABSTRACT

INTRODUCTION

Bioremediation is the process by which living organisms, generally bacteria, degrade or transform hazardous compounds into less toxic compounds [1]. At present, two classes of pollutants represent important challenges for bioremediation; one is polycyclic aromatic hydrocarbons (PAHs), and the second is halogenated hydrocarbons.

The management of PAHs is considered challenging for many reasons. Due to their hydrophobicity, these compounds tend to accumulate in soil organic matter; thus, their desorption from soil limits their availability to microorganisms for biodegradation [2]. PAHs are compounds with two or more fused benzene rings. They are formed during the incomplete thermal combustion of solid and liquid fuels or are derived from high-temperature (500–800 °C) industrial activities or from the injection of organic materials, including coal tars, crude oil and petroleum products [3], at temperatures below 300 °C. Moreover, nautical vessel effluents and spills produce serious aquatic pollution. As a class, PAHs are relatively unreactive chemically, with low solubility in water, high melting and boiling points and low vapour pressure [4]. They are ubiquitously present, toxic contaminants. PAHs are particularly good substrates for the cytochrome P-450 found in mammalian livers, where they are converted into epoxides that may bind to DNA [5]. These epoxides, particularly those derived from PAHs with exposed "bay regions" (e.g., chrysene), are highly potent xenobiotics and suspected [6] and [7] of being mutagenic and carcinogenic [8].

PAHs released into the environment could volatilise, photo-oxidise, chemically oxidise, bioaccumulate, or adsorb onto soil [9], [10] and [11].

The bioremediation of these molecules is typically achieved by using bacteria to degrade them. Although this approach is very efficient and costs little, it is often limited by environmental conditions such as pH temperature and metal ions and salt may produce unwanted or toxic products. Despite the massive number of publications on bioremediation (more than 25,000 over the past 3 years, according to the ACS database), indicating great research interest worldwide, only a few (less than 1%) publications have focused on the chemical aspects thereof, using either a computational or multidisciplinary approach to study PAH. Given this broad background, it is difficult to focus attention on the very important aspect of enzymatic bioremediation because often the results are not straightforward. Thus, this work aims to highlight only those articles that play a key role in the search for enzymes, conditions or methods applicable to bioremediation.

This review is divided into two main chapters. The first addresses enzymes with a well-resolved 3D structure (X-ray or NMR data) that can degrade PAH, with some experimental details regarding new or lesser known enzymes provided; the second chapter addresses data and information on the computational aspects of enzymatic bioremediation.

ENZYMES DEGRADING POLYCYCLIC AROMATIC HYDROCARBONS

In an effort to remove as many PAHs as possible in the shortest time, many researchers are trying to discover new enzymes or strands in wild-type bacteria with the aim of engineering or immobilising them on opportune devices for bioremediation. This research has two aspects, the first being the search for new bacteria or enzymes able to degrade PAHs over a broad spectrum of chemical conditions. Table 1 reports the results obtained by Arun et al. [1] regarding various bacteria and conversion efficiencies; interestingly, as shown, when bacteria are mixed, the conversion rates decrease. The second aspect is the attempt to develop experimental conditions or molecular stand-ins as mediators to increase the degradation efficiency of target molecules. In particular, Arun's [1] work clearly notes the need for more targeted studies on enzyme bioremediation and any external factors that influence it. For example, it is not clear why mixing enzymes in some cases greatly reduces PAH removal efficiencies but in others has no effect. For examplePseudomonas sp. (see Table 1) can degrade 95% of available pyrene and Pleurotus ostreatus can degrade 32% of pyrene, yet by mixing the two bacteria, only 17% pyrene is degraded. The study by Arun et al. highlights how triggering chemical parameters may alter the degradation activity of bacteria; in particular, it is expected that the mixture of more than one type of bacterium affects the degradation of small molecules, ions or enzyme–enzyme interactions, as reported by

Vinas et al. [12]. Vinas et al. studied a highly creosote-contaminated soil, observing that the addition of nutrients, moisture content and aeration were the key factors of PAH bioremediation, as reported in Table 1; however, it was evidenced that there was a remarkable difference in the composition of the bacterial community. Moreover, there are other parameters that could be necessary to consider, such interspecies interactions, nutrient effects, changes in PAH bioavailability and recalcitrant effects. Along with performing studies on bacteria, strain types and mixtures thereof, many authors have tried to study and characterise new enzymes.

Table 1: Degrading activity of some simple bacteria, and in mixture [1] and [12]

Organism	PAH degradation (%), standard deviation it is ≤3%				
	Naphthalene	Acenaph-thene	Fluorene	Anthra-cene	Pyrene
Pseudomonas sp.	15.5	28.0	24.4	25.4	92.3
Pycnoporus sanguineus	12.0	7.0	17.6	15.6	4.4
Coriolus versicolor	27.4	2.0	23.0	22.4	42.0
Pleurotus ostreatus	29.4	20.6	20.6	19.0	32.0
Fomitopsis palustris	19.5	7.5	7.0	31.7	7.3
Daedalea elegans	35.8	5.9	5.9	2.4	26.1
Pycnoporus sanguineus mixed with Pseudomonas sp.	13.5	29	24.2	11.4	17.4
Coriolus versicolor mixed with Pseudomonas sp.	15.5	27	24	25.0	93.7

Pleurotus ostreatus mixed with Pseudomonas sp.	13	25	19	20.0	17.0
Fomitopsis palustris mixed with Pseudomonas sp.	13.1	16.3	16.3	12.0	93.7
Daedalea elegan mixed with Pseudomonas sp.	23	14.9	14.9	3.4	46.4
Aerated soil at 40% WHC in presence of Sphingomonas and Azospirillum		100	100	84	87
Aerated soil at 40% WHC; KNO_3 and K_2HPO_4 in presence of Sphingomonas and Azospirillum		100	100	81	90
Aerated soil at 40% WHC; nutrients; biosurfactant MAT10		100	100	79	90
Aerated soil at 40% WHC; nutrients; ferric ion added as ferric octoate		100	100	87	88

Environmental and Chemico-physical Parameters Affecting Degradation

A pivotal aspect of enzymatic degradation that in many cases can explain abnormal efficiency, reaction kinetics or selectivity is represented by chemico-physical parameters and other variables affecting PAH degradation [13]. One of these variables is the aerobic condition of the reaction; in particular, it is well known that the removal of PAHs under anaerobic conditions is normally two-fold less efficient than that under aerobic conditions (Table 2). However, McNally et al. shed light on PAH bioremediation also being possible under severe conditions such as anaerobic conditions [14]. McNally reports three pseudomonad strains with activities, which are in some cases analogous to aerobic strains. The time required for the colonisation of a culture medium by the appropriate degradation organism is critical, but once this step is reached, PAH conversion is as quick as that under aerobic conditions. Though aerobic conditions appear to be pivotal for PAH removal, slight changes in structure may overcome external degrading conditions. This idea was developed by Shuler et al. [15]. Shule et al. discovered a new pool of strains fromSphingomonas sp. that can oxidise low-molecular-weight PAHs, chlorinated biphenyls, dibenzo-p-dioxin and high-molecular-weight PAHs such as benz(a) anthracene, chrysene and pyrene. The authors hypothesise that a pool of highly conserved multi-component dioxygenases that exhibit slight structural variations in their amino acid sequences outside the catalytic pocket may appear to be responsible for larges differences in selectivity towards PAHs. Another chemical parameter with a degradation-modulating effect was reported by Mancini et al. [16], who suggested that PAH degradation may be modulated by trace iron elements [17]. The authors conducted experiments on the modulation of iron ion concentrations ([Fe]/ [Toluene] = 10-1 and 10-3), optimising the corresponding isotope analysis to obtain evidence of the significant effect of these trace elements; however, it is clear that further study is required (Table 2). One of Mancini's experiments shows that the trace element cobalt is required to drive the reductive dechlorination of chlorinated ethenes by vitamin B12; hence, examining the effects of cobalt limitations on enzymatic activity and isotopic fractionation is warranted. The possible mechanism governing the activity of iron ions towards PAH degradation was studied by Santos et al. [18]. The researchers established that iron ions enhance anthracene degradation directly by increasing the activity of the enzymes involved in the aerobic biodegradation (Table 2) pathways of hydrocarbons and indirectly by increasing PAH solubility due to stimulation through biosurfactant production [14] and [19]. Furthermore, Santos et al. noted that iron ion activity directly correlates with salt solubility, Fe_2O_3 being less active than $Fe(NO_3)_3$. Although the extent of degradation increases, it is not accompanied by a significant change in the degradation rate, indicating a new possible degradation pathway. It is noted that the proteins in the active site of the degrading enzyme usually contain several iron ions, such proteins being commonly categorised as non-heme iron proteins.

These enzymes form a broad class of molecules that exhibit a Rieske centre. The presence of iron ions at the active site of this class of protein supports the hypothesis that external iron ions may be involved in enzymatic activity in an unknown way. If this hypothesis is confirmed by other studies, the iron ion may be recognised as a crucial factor in bioremediation, on the one hand increasing the solubility of recalcitrant PAH and on the other opening the possibility of hitherto unexplored degradation pathways. Another parameter that is critical to the efficiency of many degradation reactions is pH level. In particular, Sood and Lal [20] have revealed novel yeast species isolated from soil samples contaminated with acidic oily sludge (pH 1–3) that can degrade 73% of total petroleum hydrocarbons at pH 3 within a week. In this case, an enzyme with 60% homology to cytochrome P-450, one of the first enzymes studied for bioremediation [21] and [22]. Again, environmental conditions (pH) determine enzyme activity. Additionally, the author discovered that this new strain could function under worse conditions typical of real applications; thus, more studies are required to better understand the defense mechanisms of this strain against pH. In many of the cases reported herein, the first step of degradation is the vehiculation of PAH to the enzyme or bacterium. This step represents, in most cases, the slowest step. A recent study in this direction was reported by Castelli et al. [23] and [24], who used calorimetry to study degradation. In particular, the researchers discovered that PAHs could interact with model membranes but were unable to migrate through an aqueous medium to reach biological membranes (Table 2). Furthermore, PAHs can be transferred from loaded vesicles to empty large unilamellar vesicles (LUVs). These results suggest that lipophilic agents favour absorption such that these interactions should correlate with other mechanisms caused by the transfer from a lipophilic medium to a biological membrane. The possibility of using LUVs as carriers should obviate this transfer, leading to a great increase in PAH degradation. Another attempt to overcome the solubilisation step of PAH degradation was reported by Eibes et al. [25]. The authors tried to enhance the solubility of PAHs by using an acetone solution (36%, v/v). The results clearly demonstrate great extents of degradation (greater than 95%) for anthracene, dibenzothiophene and pyrene achieved in less than 24 h (Table 2).

Moreover, the authors shed light on these degradation mechanisms by analysing degradation residues. The study highlighted an undocumented mechanism for dibenzothiophene (Fig. 1panel b) with an intermediate that breaks PAH into a simpler acid and pyrene (Fig. 1 panel a). In addition to the studies already discussed, an attempt to understand and increase the solubility of PAHs in an aqueous environment (Table 2) was reported in a recent paper by Vacha et al. [26]. The authors investigated the adsorption of benzene, naphthalene, anthracene and phenanthrene at the water interface using molecular dynamics simulations. In that paper, MD simulations were employed to reconstruct the potential mean force (PMF) profile for the process of transforming from a gas phase to an aqueous one. The results indicate that, in all cases, a deep minimum in the free energy profile corresponding to the water-gas interface occurs. This finding points to the importance of the aqueous surface for the chemistry of PAHs. The free energy minima of PAH molecules at the air–water interface imply that in cases in which the surface area is large, the surface reactivity of PAH molecules can be more significant than the bulk chemistry, such as in atmospheric droplets, ice, snow and thin water films on aerosols.

Table 2: Chemical parameters influencing PAH degradation

Enzyme or strain or molecule studied	Parameter	Effect	Reference
Pseudomonas	Aerobic condition	Anaerobic conditions is normally two-fold less efficient than in aerobic conditions	McNally et al. [14]
Pseudomonas putida	Iron ions	Kinetic of degradation	Mancini et al. [16]
Pseudomonas sp.	Iron ions	Conversion rate	Santos et al. [18]
Laccase	Presence of: vanillin, acetovanillone, acetosyringone, syringaldehyde, 2,4,6-trimethylphenol and p-coumaric acid	Conversion rate	Canas et al. [35]
Cytocrome P450	pH < 3	Conversion rate	Sood and Lal [20]
PAH only	Solubility	Conversion rate	Castelli et al. [24]
PAH only	Solubility	Conversion rate	Eibes et al. [25]
PAH only	Medium composition and solubility	Conversion rate	Vinas et al. [12]
Laccase	Presence of: 2,2'-azino-bis-(3-ethylbenzothiazoline-6-sulphonic acid, diammonium salt (ABTS) and 1-hydroxybenzotriazole HBT)	Conversion rate	Farnet et al. [31]

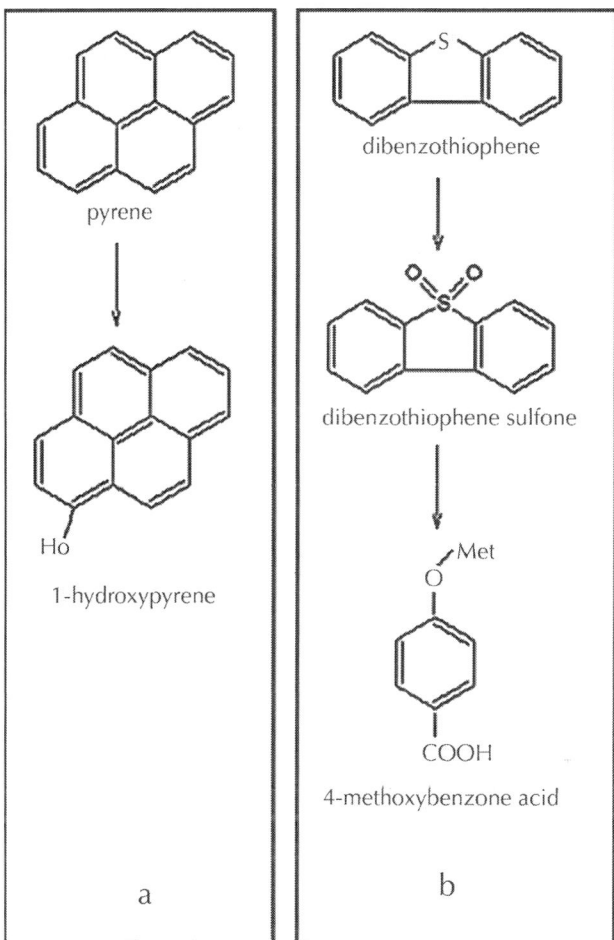

Figure 1: Pyrene (a), dibenzothiophene (b) and their oxidation products [26].

Manganese Peroxidase

In a study on highly efficient degrading enzymes, Hofrichter et al. [27] investigated manganese peroxidase (MnP) from Nematoloma frowardii to shed light on the concept of enzymatic combustion [28], discovering a molecule that is able to degrade a broad range of pollutants such as 2,4,6-trinitrotoluene and catechol. Despite this degrading ability, MnP is unable to degrade some PAHs such as pyrene and in addition, its degradation mechanism is not well understood. Notwithstanding these difficulties, the availability of the well-resolved structure of MnP makes this enzyme attractive for applied research in this field. In fact, many studies focus on exploiting enzyme mutation [29] to increase the efficiency of MnP and better understand PAH-enzyme interactions. Zang et al. obtained some mutants of MnP by applying site-directed mutagenesis to arg42 and ASN131. The results revealed some species with different degrading activities, but the lack of accurate engineering strategies and characterisation of mutants has prevented researchers from obtaining information that is more useful. Although the work of Zang is one of the first efforts in this meaningful direction, other researchers have explored the route of immobilisation, which represents the first step in the search for versatile applications of this enzyme. Acevedo et al. [30] obtained immobilised enzymes with higher efficiency than free enzymes. After 24 h, manganese peroxidase (MnP) immobilised on nano-clay efficiently transformed anthracene and pyrene into anthraquinone and 4,5-dihydropyrene, respectively and, to a lesser extent, fluoranthene and phenanthrene. Immobilised MnP was generally twice as efficient as free MnP. As evidenced in this review, Acevedo drew attention to the environmental factors affecting degradation.

Laccase

Farnet et al. focused on six different isoforms of laccase from Marasmius quercophilus [31] that were able to oxidise widespread pollutants, such as PAHs. Though many crystallographic structures of laccase are available, Farnet et al. lacked this information in their study, but their work is reported in this study because it provides useful hints regarding degradation that we believe may be applied to other degrading enzymes. In vitro studies confirmed that these laccase isoforms are able to transform anthracene and benzo(a)pyrene, though naphthalene and phenanthrene were not degraded. In the same work, Farnet et al. noted that the conversion of anthracene to anthraquinones may be greatly enhanced by using 2,2'-azino-bis-(3-ethylbenzothiazoline-6-sulphonic acid) diammonium salt (ABTS [32]) and 1-hydroxybenzotriazole (HBT), which act as electron transporters for laccase. Furthermore, this study revealed that the determining factor for oxidation was the ionisation potential (IP); indeed, only molecules with an IP < 7.55 eV was degraded. Unfortunately, though dozens of laccase structures are currently known, no 3D structures of this specific enzyme have been investigated. Although it is remarkable that enzymes of the same class may have exhibit different responses to degradation, an approach based on 3D structure may lead to the possibility of engineering enzymes that are specifically designed for bioremediation. As reported for MnP, immobilisation represents the first step in the commercialisation of enzymes and some authors are currently pursuing this step. However, although laccases are promising enzymes [33] and [34] for degrading PAHs, their efficiency decreases rapidly during the industrialised process of immobilisation. In a recent study, Hu [33] immobilised one isoform of laccase from Trametes versicolor on silica nanoparticles with the aim of developing an efficient industrial application for this system such that when it was immobilised it became less active due to partial unfolding or underwent general structural destabilisation. In many studies, it has been demonstrated that immobilised enzymes are able to degrade anthracene (ANT) to anthraquinone (ANQ) via a pH-dependent mechanism. Considering the evident differences in efficiency between laccase and MnP, as reported above, a comparative study of the different enzyme–substrate interactions should prove interesting, with the objective of minimising the altered functions of the immobilised enzymes. Moreover, the concept of using a mediator in enzyme degradation may be relevant to industrial applications, especially for minimising costs. Indeed, mediators may become crucially important in future studies on the mechanisms involved in the use of mediator activities as tools for optimising enzyme activity. In a recent work by Canas et al., the authors provided evidence that chemical conditions are very important in determining PAH degradation, particularly for laccase [35]. Such enzymes have an in vitro conversion rate of approximately 15%, but with other molecules such as vanillin, acetovanillone, acetosyringone, syringaldehyde, 2,4,6-trimethylphenol and p-coumaric acid, their activity is greatly enhanced (Fig. 2). The mechanism of mediated oxidation is not yet clear, although different routes are believed to be involved depending on the mediator involved [36].

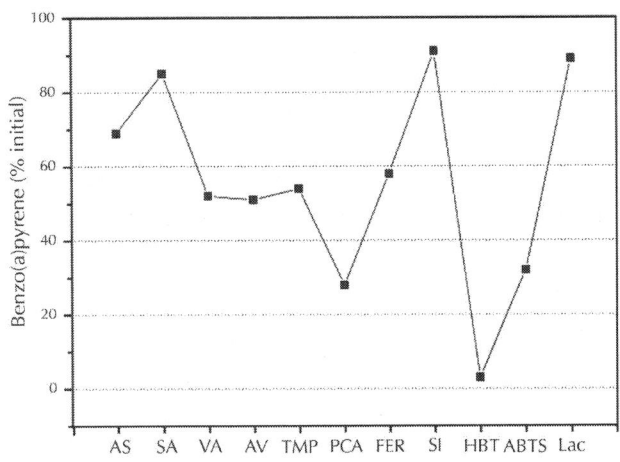

Figure 2: Residual benzo[a]pyrene (in % of initial 50 M) after 24-h oxidation with laccase and different mediators (500 M). Mean values and 95% confidence limits are shown. Mediators: acetosyringone (AS), syringaldehyde (SA), vanillin (VA), acetovanillone (AV), 2,4,6-trimethylphenol (TMP), p-coumaric acid (PCA), ferulic acid (FER), sinapic acid (SI), HBT, and ABTS (Lac, laccase without mediator) [36].

Soybean Peroxidase

The same observations discussed above have been reported by other authors in studying an enzyme that has long been known to be involved in PAH degradation, soybean peroxidase (SBP). SBP catalyses the oxidation of a variety of PAHs but only if the pollutants are dissolved in organic solvents such as acetonitrile, tetrahydrofuran, or dimethyl formamide. Another shortcoming of SBP [37] is the limited pH range (2–2.5) over which it functions efficiently. Within this range, the conversion rate of SBP is very high (>90%) and despite its limitations, this enzyme is a good choice because of its efficiency. There have been many attempts to overcome the limitations of free SBP [38] and [39] but to no avail; thus, further study is required. Nevertheless, SBP immobilisation is a well-known and efficient degradation route.

Naphthalene Dioxygenase

To date, the only PAH dioxygenase used in industrial applications is naphthalene dioxygenase (NDO) [40]. McIver reports that NDO is well suited to producing intermediates for the chemical industry; for example, diols produced from PAH degradation represent an ideal starting point for many chemical syntheses. Moreover, NDO is very interesting because of its degradation of a broad range of pollutants and it was one of the first molecules discovered to initiate bioremediation naturally. In many cases, researchers [40],[41] and [42] have adopted NDO not only as a degradation tool but also as a biomarker. NDO activity is considered a precursor to natural biodegradation. All of the enzymes herein listed – MNP, SBP, NDO, Laccase and AKR – are bioremediable, but much work is needed to investigate how they operate in a complex matrix that is both biological and chemical. To better steer further studies, we collected significant data in Table 3; all selected enzymes have a known 3D structure, which is key to deeply understanding all of their characteristics and fundamental for further engineering the enzymes.

Table 3: Data for manganese peroxidase (MnP), soybean peroxidase (SbP), naphthalene dioxygenase (NDO), laccase

Enzyme	PDB id.	Degradation capability	Immobilisation	Reference
MnP	1MNP	2,4,6-Trinitrotoluene and catechol but not pyrene	Degrade anthracene and pyrene	Kirk and Farrell[28], Acevedo et al. [30]
SbP	1FHF	Degrade a broad range of pollutants including PAH	–	Kraus et al. [37]

NDO	1O7H, 1O7G, 1O7 N, 1O7P, 1O7 W	Degrade a broad range of pollutants	–	Di Gennaro et al. [42]Wammer et al. [54]
Laccase	3KW7, 3FU7, 3FU8, 3FU9, 3DIV, 2ZWN, 3F8X, 3CG8, 4A2D, 4A2E, 4A2H, 2Q9o, 2QT6, 2HRG, 2HRH, 2H5U, 2IH8, 2IH9, 1V10, 1GYC, 1GW0, 1KYA, 1HFU, 1A65, 1UVW	Anthracene and benzo(a)pyrene but not naphthalene and phenanthrene	Degrade anthracene	Farnet et al. [31] Hu et al. [33]

POLYCYCLIC AROMATIC HYDROCARBONS: COMPUTATIONAL ASPECTS

Bioinformatics Methods

Studies on laccase have been directed towards the determination of the kinetics of degradation. Cristovao et al. [43] and [44] used a mathematical approach to determine the kinetic constants that adequately describe the degradation kinetics of some reactive textile dyes. The results were confirmed by comparing the time courses obtained experimentally with those obtained from the model the authors developed. These results may be used to predict the time courses of substrate consumption and product formation under different substrate concentrations. Thus, establishing kinetic models for these reactions is a useful tool for simulating and designing enzymatic bioreactors. Interestingly, the kinetic data calculated by the models coincide well with the experimental data. Although this work is not directly related to enzyme bioremediation, the method used to calculate kinetic data may be extended to other systems. While the techniques employed in such studies appear to be valid as a methodological approach and the results thus obtained are in good agreement with experimental data, we believe that a more in-depth approach should be applied to enzymatic systems to enhance our knowledge of enzyme–PAH interactions. Other bioinformatics methods, particularly electrostatic methods and experiments, were applied by Brown et al[45] to determine the redox potential of NDO. The large range of reduction potentials for Rieske ferredoxins (from −150 to +400 mV) was first suspected to arise from the different extents of solvent accessibility to the cluster, but studies performed to determine the structure of Rieske ferredoxin proteins and related studies have demonstrated that there are no large variations in solvent accessibility. Subsequent observations indicate that differences in the electrostatic environment and not structural differences between Rieske proteins are responsible for the wide range of reduction potentials observed. Brown et al. developed a model to predict the reduction potential of Rieske proteins given only their crystal structure. The method proposed accurately predicts the reduction potentials of 17 Rieske proteins for which both the structures and experimentally determined potentials are available. Additionally, the method employed has a bright future in facilitating in silico prediction of the effects of mutations on Rieske protein reduction potentials and estimating reduction potentials for newly determined Rieske ferredoxin structures.

Docking Calculations

A docking approach for laccase was proposed by Suresh et al. The researchers analysed the binding of laccase to a broad range of molecules to develop a useful tool for finding putative pollutants for other biodegrading enzymes [46]. The authors observed a good match between the predicted bioremediation of laccase and the results of experiments on anthracene and phenanthrene degradation (Table 4). The authors reported that, in some docking simulations, the experimental data diverged because of the scoring function used and metal-related problems in docking, which are further complicated by the difficulty of reproducing the multiple coordination geometries of the copper complex [47] and [48]. NDO was also studied using a docking approach. Carredano et al. [49] docked low-molecular-weight PAHs (indole, naphthalene and biphenyl) with NDO and the results of their study suggest the presence of pockets reserved for the binding of the aromatic ring. The probable binding site of dioxygen is located between this pocket and the catalytic iron. A similar approach using the same enzyme was adopted by Librando and Forte, who employed molecular dynamics and docking techniques to explore new structures similar to wild-type NDO [50]. The authors created a library of opportune fragments in silico and carried out MD and docking simulation. The simulation results for these fragments, accounting principally for energy parameters, produced a short list of peptides with strong binding activity, which may be a great boon for future laboratory work. Further field studies of the same group [51] will increase the accuracy of the binding parameter and help generate a larger library by adopting hydrophobicity, free binding energy and RMSD as indicators to better highlight binding zones and modifications that are able to help NDO bind PAH efficiently. Moreover, this field of research generally utilises a database approach, but in this study, a database had to be generated specifically for the target enzyme. Such studies offer interesting opportunities for molecular screening to highlight active strands against PAH but also evidence the need to manage large databases of molecules and fragments, as addressed in the final part of this review. The first step in the biodegradation of aromatic hydrocarbons often involves the dihydroxylation of two adjacent carbon atoms on the aromatic ring, catalysed by ring-hydroxylating dioxygenase (RHD). Several bacteria have been found to degrade PAHs, but only a few have been reported to attack four- and five-ring PAHs [52], [53] and [54]. Jakoncic et al. [55] indicate that the broad substrate specificity of the dioxygenase fromSphingomonas sp. Strain CHY-1i (PhnI) [56] is primarily due to the large volume and particular shape of the enzyme's catalytic pocket. Molecular simulations of the PhnI pocket revealed the pocket to be at least 2 Å longer and wider at the entrance, a unique feature of dioxygenases with known structure that certainly allows five-ring benzo(a)pyrene to bind to catalytic Fe. Modelling various PAHs shows that Phe 350 in the central region of the pocket is essential for regio- and substrate-specificity, whereas Leu 223 and Ile 260 in the distal region provide the specificity of high-molecular-weight PAHs. Further studies involving replacements for the specific residues of substrate-binding pockets by site-directed mutagenesis should provide new insight into the role of these residues in the catalytic activity of the enzyme. Along the same direction of research, the authors of the present review adopted similar strategies and extended their potential in targeting some amino acids that are interesting for future mutagenesis and indicate a shape factor in engineering enzymes that are able to degrade PAH [57] and [58].

Table 4: GOLD average fitness scores for known substrates, a few predicted targets and newly predicted targets for bioremediation [47]

S. no.	Name	GOLD average fitness score	
		Trametes versicolor	Bacillus subtilis
1	ABTS	50.58	48.14
2	Anthracene	40.37	30.22
3	Phenanthrene	42.05	31.62
4	Thiodicarb	59.01	41.61
5	Malathion	57.29	48
6	Captan	44.23	39.27
7	Atrazine	44.24	30.29

8	Indigo	44.6	40.34
9	Remazol red B	47	33.5
10	Vanilic acid	31.86	–
11	2,4-Dichlorophenol	30.22	30.66
12	m-Chlorophenol	30.25	–
13	2,4,6-Trichlorophenol	32.17	31.94
14	Sinapic acid	37.67	–
15	Syringadazine	33.32	30.3

Quantum Mechanical Techniques

One of the widest and most complete works on enzyme degradation was reported in 2006 by Wammer et al. [54]. The authors systematically collected PAHs able to interact with NDO and the reaction products thereof. The most interesting part of this work was the application of in silico bioremediation via DFT studies and modelling, the results of which indicate that thermodynamically all PAHs can interact with the active sites of NDO. For both enzymes and PAHs, only steric hindrance determines which molecules can efficiently react with NDO. To better comprehend PAH-enzyme interactions, a direct application of the structure/reactivity relationship was adopted by Librando and Alparone [59]. A mixture of ab initio and functional density theory calculations for a series of dimethylnaphthalene (DMN) isomers was adopted to predict the PAH degradation efficiency. The results support the idea that electronic polarisability may be a useful tool for predicting the biodegradation trends of a series of compounds, besides playing a fundamental role in the biodegradation process of DMNs and providing a theoretical basis for Farnet's hypothesis discussing at the beginning of this paper. Strictly related to the structure reactivity/relationship for an enzyme, the method was combined with docking techniques to effectively characterise PAH–NDO interactions.

The mechanism of the interactions between PAH and enzymes is not yet fully understood and it is likely that it will not be easily accessed using experimental techniques. Recent studies employing a quantitative structure–activity relationship (QSAR) approach indicate that some experimental tools may offer great help in this respect. Li F. et al. [60] and [61] investigated the binding interactions between PAHs and various substrates such as DNA or oestrogen receptors. The descriptors incorporated into the QSAR models indicated that the binding activity was related to molecular size, van der Waals volume, shape profile, polarisability and electrotopological state, hydrogen bonding, hydrophobicity and p–p interactions. In those studies, QSAR was adopted to screen mutations, in silico, that improve enzyme reactivity. Moreover, QSAR has been applied to provide information about the toxicity of PAHs and any degradation intermediates [62], revealing that a significant relationship exists between toxicity and lipophilicity (K_{ow}), which suggests that non-polar narcosis is the prevalent toxic effect of the tested PAHs. This result is observed because toxicity, which is directly related to lipophilicity for biological membranes (i.e., non-polar narcosis), depends mostly on the amount of the compound accumulated in the same membranes. In addition, the ionisation potential of PAHs has been identified as an important parameter in explaining their toxic effects in terms of their log K_{ow}.

Database Approach

The enormous amount of data regarding reaction environment and degradation creates a serious problem in finding the right reaction, degradation products, etc. required. Thus, a database approach may be useful. One complication associated with the database approach is the integration of heterogeneous sources of information related to bioremediation. In this respect, the work by Pazos et al. offers a troubleshooting 'metarouter' to help in such cases [63], which is a useful instrument for assessing the environmental fate of compounds or mixtures and designing biodegradative strategies for these species. For chemical compounds, the following information is provided: name, synonyms, SMILES code, molecular weight, chemical formula, image of the chemical structure, canonical three-dimensional structure in PDB format, physicochemical properties (density, evaporation rate, melting point, boiling point and water solubility—the user can define and input new ones) and links to other databases. For reactions, the following information is provided: substrates and products, catalysing enzyme and links to other databases. For enzymes, the following information is provided: name, Enzyme Commission (EC) code, organisms where the gene is present, database sequence identifiers and links to other databases. The limitation of the metarouter is that it focuses on the biochemical aspects of biodegradation rather than the nature of the biomolecules carrying out the reactions. In a recent study, Carbajosa presented a new database, Bionemo [64], which is a resource that complements other biodegradation databases. Bionemo was built by manually associating data from published articles and, in general, from the biodegradation literature and linking them to an underlying biochemical network. Currently, Bionemo contains sequence information for 324 reactions and transcription regulation information for more than 100 promoters and 100 transcription factors. Meanwhile, current biodegradation databases link reactions to protein sequences in databases that have been annotated with the corresponding EC codes. However, this method may be inaccurate. For instance, many reactions share the same EC code, although they use distinct substrates and generate different products. The Bionemo database combines metabolic, genetic and regulatory information. The central entries of the database are enzymatic complexes. These are linked to biochemical reactions that transform substrates into products. Reactions are associated with different pathways.

Conclusions: Theoretical Methods Can Support Experiments

The objective of this review was to provide an overview of notably different problems, such as those associated with the organisation of bioremediation research, problems regarding enzyme activity, MD, docking and DFT; these problems, however, have several features in common. Bioremediation pathways are ultimately reduced to processes that remove PAHs and feature complex biological matrices and various chemical conditions such as pH, metal ions, etc. In this paper, we have presented select evidence regarding many new enzymes with known 3D structures obtained from X-ray or NMR, as well as highlighted important aspects regarding the removal of PAHs. Although experimental approaches provide highly reliable data that are not comparable to those yielded by many in silico methods, theoretical approaches offer certain advantages. For example, Carredano [49] produced mutants by site-directed mutagenesis to study the effect of individual amino acids on the degradation activities of NDO, undeniably demonstrating that the cost of such experiments is great.

Today, with the modern techniques of modelling and high-performance computing centres, such experiments could be considerably improved by performing prescreening studies, such as the study performed by the authors of this review in the case of the enzyme PhnI. Similarly to Carredano, Librando studied the effect of the mutation of single amino acids on the affinity of enzymes for some molecules of environmental interest. Of course, pairing MD with docking techniques will not replace but supplement studies on site-specific mutagenesis and experimental approaches in general by reducing the number of experiments that must be performed in vitro and consequently costs.

Other theoretical studies involving DFT and QSAR may boost current studies in the field of bioremediation by both shortening the time required to obtain new chimerical enzymes and reducing the number of laboratory experiments that must be performed. Other useful techniques that could supplement experiments include quantum mechanical (QM) techniques, which can predict the reactivity of receptor molecules and in some cases also some characteristics of substrates. Of course, these techniques can be applied only to small systems due to the high complexity of the calculations. In this context, QM methods can be adopted as methods supporting MD/docking, offering a vision of effects that would otherwise not be accessible. Finally, the use of databases makes it possible to accelerate the selection of molecules or reaction pathways through the use of intelligent algorithms. All theoretical studies regarding, for instance, kinetics, DFT, docking and molecular dynamics shed light on the great potential of in silico bioremediation. Moreover, the results of computer simulations used to study laccase agree well with corresponding experimental data, indicating a high degree of reliability (Table 5). The studies on laccase do not describe any enzyme mutations but offer interesting hints. NDO and PhnI are both good candidates for further studies and have been well characterised via in silico approaches. This work should offer a useful perspective of powerful in silico tools for incorporating structural modifications into selected enzymes.

Table 5: This table summarises theoretical calculation and experiments comparison on enzyme

Enzyme	Technique	PAH docked	Experiments comparison	Reference
Laccase	Docking	Yes	Yes	Suresh et al. [48]
Laccase	Kinetic of degradation	No	Yes	Cristovao et al. [44]
NDO	DFT/AB initio	No	No	Librando and Alparone [59]
NDO	MD/docking	Yes	No	Librando and Forte [50]
PhnI	MD/docking	Yes	Yes	Jakoncic et al. [55], Librando and Pappalardo [57], Librando and Pappalardo [58]
Estrogen receptor α (enzyme adopted for functional studies)	MD/Docking/QSAR	Yes	No	Li et al. [61]

Though it is not clear if there is a crucial factor for degradation, essential characteristics clearly include 3D structure, the presence of small molecules and chemical conditions. The present study will thus serve as an important reference in planning future experiments.

ACKNOWLEDGMENTS

This work was supported by the Italian Ministry for Research and University (MIUR), Program PRIN 2009and Cometa Consortium.

REFERENCES

1. Arun, Polycyclic aromatic hydrocarbons (PAH) biodegradation by basidiomycetes fungi, Pseudomonas isolate, and their cocultures: comparative in vivo and in silico approach, Applied Biochemistry and Biotechnology 151 (2008) 132–142.
2. G. Eibes, T. Lu-Chau, G. Feijoo, M.T. Moreira, J.M. Lema, Complete degradation of anthracene by manganese peroxidase in organic solvent mixtures, Enzyme and Microbial Technology 37 (2005) 365–372.
3. A.K. Haritash, C.P. Kaushik, Biodegradation aspects of polycyclic aromatic hydrocarbons (PAHs): a review, Journal of Hazardous Materials 169 (2009) 1–15.
4. Working Group PAH, Ambient Air Pollution by Polycyclic Aromatic Hydrocarbons (PAH), 1, Office for official publications of the European Communities, 2001, pp. 56.
5. S. Nesnow, C. Davis, W.T. Padgett, L. Adams, M. Yacopucci, L.C. King, 8,9-Dihydroxy-8,9-dihydrodibenzoa,l.pyrene is a potent morphological celltransforming agent in C3H10T(1)/(2)Cl8 mouse embryo fibroblasts in the absence of detectable stable covalent DNA adducts, Carcinogenesis 21 (2000) 1253–1257.
6. Luch, A. Glatt, H. Platt, K.L. Oesch, F. Seidel, A. Synthesis, Mutagenicity of the Diastereomeric Fjord-Region 11,12-dihydrodiol 13,14-epoxides of dibenzo a,L.pyrene, Carcinogenesis 15 (1994) 2507–2516.
7. S. Amin, D. Desai, W. Dai, R.G. Harvey, S.S. Hecht, Tumorigenicity in newborn mice of Fjord Region and other

sterically hindered diol epoxides of benzo G.chrysene, dibenzo a,L.pyrene (Dibenzo Def,P.Chrysene), 4h-cyclopenta Def.chrysene and fluoranthene, Carcinogenesis 16 (1995) 2813–2817.
8. InternationalAgency forResearchonCancer, Polynuclear aromatic compounds, part 1, chemical, environmental and experimental data, Monographs on the Evaluation of Carcinogenic Risks to Humans 32 (1983) 211–224.
9. S.R. Wild, K.C. Jones, Polynuclear aromatic-hydrocarbons in the UnitedKingdom Environment – a preliminary source inventory and budget, Environmental Pollution 88 (1995) 91–108.
10. S.Y. Yuan, S.H. Wei, B.V. Chang, Biodegradation of polycyclic aromatic hydrocarbons by a mixed culture, Chemosphere 41 (2000) 1463–1468.
11. C.J. Diblasi, H. Li, A.P. Davis, U. Ghosh, Removal and fate of Polycyclic Aromatic hydrocarbon pollutants in an urban stormwater bioretention facility, Environmental Science & Technology 43 (2009) 494–502.
12. M. Vinas, J. Sabate, M.J. Espuny, A.M. Solanas, Bacterial community dynamics and polycyclic aromatic hydrocarbon degradation during bioremediation of heavily creosote-contaminated soil, Applied and Environmental Microbiology 71 (2005) 7008–7018.
13. U.S. Obayori, S.A. Adebusoye, A.O. Adewale, G.O. Oyetibo, O.O. Oluyemi, R.A. Amokun, et al., Differential degradation of crude oil (Bonny Light) by four Pseudomonas strains, Journal of Environmental Sciences-China 21 (2009) 243–248.
14. D.L. Mcnally, J.R. Mihelcic, D.R. Lueking, Biodegradation of three- and four-ring polycyclic aromatic hydrocarbons under aerobic and denitrifying conditions, Environmental Science & Technology 32 (1998) 2633–2639.
15. L. Schuler, Y. Jouanneau, S.M.N. Chadhain, C. Meyer, M. Pouli, G.J. Zylstra, et al., Characterization of a ring-hydroxylating dioxygenase from phenanthrenedegrading Sphingomonas

sp strain LH128 able to oxidize benz a.anthracene, Applied Microbiology and Biotechnology 83 (2009) 465–475.
16. S.A. Mancini, S.K. Hirschorn, M. Elsner, G. Lacrampe-Couloume, B.E. Sleep, E.A. Edwards, et al., Effects of trace element concentration on enzyme controlled stable isotope fractionation during aerobic biodegradation of toluene, Environmental Science & Technology 40 (2006) 7675–7681.
17. N.N. Pozdnyakova, J. Rodakiewicz-Nowak, O.V. Turkovskaya, Catalytic properties of yellow laccase from Pleurotus ostreatus D1, Journal of Molecular Catalysis B: Enzymatic 30 (2004) 19–24.
18. E.C. Santos, R.J.S. Jacques, F.M. Bento, M.D.R. Peralba, P.A. Selbach, E.L.S. Sa, et al., Anthracene biodegradation and surface activity by an iron-stimulated Pseudomonas sp., Bioresource Technology 99 (2008) 2644–2649.
19. A. Gottfried, N. Singhal, R. Elliot, S. Swift, The role of salicylate and biosurfactant in inducing phenanthrene degradation in batch soil slurries, Applied Microbiology and Biotechnology 86 (2010) 1563–1571.
20. N. Sood, B. Lal, Isolation of a novel yeast strain Candida digboiensis TERI ASN6 capable of degrading petroleum hydrocarbons in acidic conditions, Journal of Environmental Management 90 (2009) 1728–1736.
21. S. Masaphy, D. Levanon, Y. Henis, K. Venkateswarlu, S.L. Kelly, Evidence for cytochrome P-450 and P-450-mediated benzo(a)pyrene hydroxylation in the white rot fungus Phanerochaete chrysosporium, FEMS Microbiology Letters 135 (1996) 51–55.
22. J. Gonzalez, N. Marchand-Geneste, J.L. Giraudel, T. Shimada, Docking and QSAR comparative studies of polycyclic aromatic hydrocarbons and other procarcinogen interactions with cytochromes P450 1A1 and 1B1, SAR and QSAR in Environmental Research 23 (2012) 87–109.
23. F. Castelli, D. Micieli, S. Ottimo, Z. Minniti, M.G. Sarpietro, V. Librando, Absorption of nitro-polycyclic aromatic

hydrocarbons by biomembrane models: effect of the medium lipophilicity, Chemosphere 73 (2008) 1108–1114.

24. F. Castelli, V. Librando, M.G. Sarpietro, Calorimetric approach of the interaction and absorption of polycyclic aromatic hydrocarbons with model membranes, Environmental Science & Technology 36 (2002) 2717–2723.

25. G. Eibes, T. Cajthaml, M.T. Moreira, G. Feijoo, J.M. Lema, Enzymatic degradation of anthracene, dibenzothiophene and pyrene by manganese peroxidase in media containing acetone, Chemosphere 64 (2006) 408–414.

26. R. Vacha, P. Jungwirth, j. Chen, K. Valsaraj, Adsorption of polycyclic aromatic hydrocarbons at the air–water interface: molecular dynamics simulations and experimental atmospheric observations, Physical Chemistry Chemical Physics 8 (2006) 4461–4467.

27. M. Hofrichter, K. Scheibner, I. Schneegass, W. Fritsche, Enzymatic combustion of aromatic andaliphatic compounds bymanganeseperoxidase fromNematoloma frowardii, Applied and Environmental Microbiology 64 (1998) 399–404.

28. T.K. Kirk, R.L. Farrell, Enzymatic combustion: the microbial degradation of lignin, Annual Review of Microbiology 41 (1987) 465–505.

29. X.M. Zhang, Y.X. Wang, L.S. Wang, G.J. Chen, W.F. Liu, P.J. Gao, Site-directed mutagenesis of manganese peroxidase from Phanerochaete chrysosporium in an in vitro expression system, Journal of Biotechnology 139 (2009) 176–178.

30. F. Acevedo, L. Pizzul, M.D. Castillo, M.E. Gonzalez, M. Cea, L. Gianfreda, et al., Degradation of polycyclic aromatic hydrocarbons by free and nanoclay-immobilized manganese peroxidase from Anthracophyllum discolor, Chemosphere 80 (2010) 271–278.

31. A.M. Farnet, G. Gil, F. Ruaudel, A.C. Chevremont, E. Ferre, Polycyclic aromatic hydrocarbon transformation with laccases of a white-rot fungus isolated from a Mediterranean schlerophyllous litter, Geoderma 149 (2009) 267–271.

32. M.T. Cambria, Z. Minniti, V. Librando, A. Cambria, Degradation of polycyclic aromatic hydrocarbons by Rigidoporus lignosus and its laccase in the presence of redox mediators, Applied Biochemistry and Biotechnology 149 (2008) 1–8.
33. X.K. Hu, P. Wang, H.M. Hwang, Oxidation of anthracene by immobilized laccase from Trametes versicolor, Bioresource Technology 100 (2009) 4963–4968.
34. A.W.J.W. Tepper, S. Milikisyants, S. Sottini, E. Vijgenboom, E.J.J. Groenen, G.W. Canters, Identification of a radical intermediate in the enzymatic reduction of oxygen by a small laccase, Journal of the American Chemical Society 131 (2009) 11680±.
35. A.I. Canas, M. Alcalde, F. Plou, M.J. Martinez, A.T. Martinez, S. Camarero, Transformation of polycyclic aromatic hydrocarbons by laccase is strongly enhanced by phenolic compounds present in soil, Environmental Science & Technology 41 (2007) 2964–2971.
36. R. Boopathy, Factors limiting bioremediation technologies, Bioresource Technology 74 (2000) 63–67.
37. J.J. Kraus, I.Z. Munir, J.P. McEldoon, D.S. Clark, J.S. Dordick, Oxidation of polycyclic aromatic hydrocarbons catalyzed by soybean peroxidase, Applied Biochemistry and Biotechnology 80 (1999) 221–230.
38. M. Nissum, C.B. Schiodt, K.G. Welinder, Reactions of soybean peroxidase and hydrogen peroxide pH 2.4–12.0, and veratryl alcohol at pH 2.4, Biochimica Et Biophysica Acta: Protein Structure and Molecular Enzymology 1545 (2001) 339–348.
39. A. Muratova, S. Golubev, L. Wittenmayer, T. Dmitrieva, A. Bondarenkova, F. Hirche, et al., Effect of the polycyclic aromatic hydrocarbon phenanthrene on root exudation of Sorghum bicolor (L.) Moench, Environmental and Experimental Botany 66 (2009) 514–521.
40. A.M. McIver, S.V.B.J. Garikipati, K.S. Bankole, M. Gyamerah, T.L. Peeples, Microbial oxidation of naphthalene to cis-1,2-naphthalene dihydrodiol using naphthalene dioxygenase in

biphasic media, Biotechnology Progress 24 (2008) 593–598.
41. J.M. DeBruyn, T.J. Mead, S.W. Wilhelm, G.S. Sayler, PAH biodegradative genotypes in Lake Erie Sediments: evidence for broad geographical distribution of pyrene-degrading mycobacteria, Environmental Science & Technology 43 (2009) 3467–3473.
42. P. Di Gennaro, B. Moreno, E. Annoni, S. Garcia-Rodriguez, G. Bestetti, E. Benitez, Dynamic changes in bacterial community structure and in naphthalene dioxygenase expression in vermicompost-amended PAH-contaminated soils, Journal of Hazardous Materials 172 (2009) 1464–1469.
43. A. Kandelbauer, A. Erlacher, A. Cavaco-Paulo, G.M. Guebitz, Laccase-catalyzed decolorization of the synthetic azo-dye Diamond Black PV 200 and of some structurally related derivatives, Biocatalysis and Biotransformation 22 (2004) 331–339.
44. R.O. Cristovao, A.P.M. Tavares, A.S. Ribeiro, J.M. Loureiro, R.A.R. Boaventura, E.A. Macedo, Kinetic modelling and simulation of laccase catalyzed degradation of reactive textile dyes, Bioresource Technology 99 (2008) 4768–4774.
45. E.N. Brown, R. Friemann, A. Karlsson, J.V. Parales, M.M.J. Couture, L.D. Eltis, et al., Determining Rieske cluster reduction potentials, Journal of Biological Inorganic Chemistry 13 (2008) 1301–1313.
46. Suresh, P.S. Kumar, A. Kumar, R. Singh, V.P. An, Insilco approach to bioremediation: laccase as a case study, Journal of Molecular Graphics & Modelling 26 (2008) 845–849.
47. X. Hu, S. Balaz, W.H. Shelver, A practical approach to docking of zinc metalloproteinase inhibitors, Journal of Molecular Graphics & Modelling 22 (2004) 293–307.
48. A. Oda, K. Tsuchida, T. Takakura, N. Yamaotsu, S. Hirono, Comparison of consensus scoring strategies for evaluating computational models of protein–ligand complexes, Journal of Chemical Information and Modeling 46 (2006) 380–391.
49. E. Carredano, A. Karlsson, B. Kauppi, D. Choudhury, R.E.

Parales, J.V. Parales, et al., Substrate binding site of naphthalene 1,2-dioxygenase: functional implications of indole binding, Journal of Molecular Biology 296 (2000) 701–712.

50. V. Librando, S. Forte, Computer evaluation of protein segments removal effects from naphthalene 1,2-dioxygenase enzyme on polycyclic aromatic hydrocarbons interaction, Biochemical Engineering Journal 27 (2005) 161–166.

51. V. Librando, A. Cambria, A. Alparone, D. Gullotto, Computational analyses of virtual proteolytic fragments generated by naphthalene 1,2-dioxygenase. In search of native-like conformation and function, Molecular Simulation 33 (2007) 231–237.

52. A.L. Juhasz, R. Naidu, Bioremediation of high molecular weight polycyclic aromatic hydrocarbons: a review of the microbial degradation of benzo a.pyrene, International Biodeterioration & Biodegradation 45 (2000) 57–88.

53. R.A. Kanaly, S. Harayama, Biodegradation of high-molecular-weight polycyclic aromatic hydrocarbons by bacteria, Journal of Bacteriology 182 (2000) 2059–2067.

54. K.H. Wammer, C.A. Peters, Polycyclic aromatic hydrocarbon biodegradation rates: a structure-based study, Environmental Science & Technology 39 (2005) 2571–2578.

55. J. Jakoncic, Y. Jouanneau, C. Meyer, V. Stojanoff, The catalytic pocket of the ring-hydroxylating dioxygenase from Sphingomonas CHY-1, Biochemical and Biophysical Research Communications 352 (2007) 861–866.

56. Y. Jouanneau, C. Meyer, J. Jakoncic, V. Stojanoff, J. Gaillard, Characterization of a naphthalene dioxygenase endowed with an exceptionally broad substrate specificity toward polycyclic aromatic hydrocarbons, Biochemistry 45 (2006) 12380–12391.

57. V. Librando, M. Pappalardo, Computational study on the interaction of a ringhydroxylating dioxygenase from Sphingomonas CHY-1 with PAHs, Journal of Molecular Graphics & Modelling 29 (2011) 915–919.

58. V. Librando, M. Pappalardo, Engineered enzyme interactions with polycyclic aromatic hydrocarbons: a theoretical approach, Journal of Molecular Graphics and Modelling 36 (2012) 30–35.
59. V. Librando, A. Alparone, Electoonic polarizability as a predictor of biodegradation rates of dimethylnaphthalenes. An ab initio and density functional theory study, Environmental Science & Technology 41 (2007) 1646–1652.
60. F. Li, X. Liu, L. Zhang, L. You, J. Zhao, H. Wu, Noncovalent interactions between hydroxylated polycyclic aromatic hydrocarbon and DNA. Molecular docking and QSAR study, Environmental Toxicology and Pharmacology 32 (2011) 373–381.
61. F. Li, H. Wu, L. Li, J. Zhao, W.J.G.M. Peijnenburg, Docking and QSAR study on the binding interactions between polycyclic aromatic hydrocarbons and estrogen receptor, Ecotoxicology and Environmental Safety 80 (2012) 273–279.
62. K. Kobeticova, Z. Simek, J. Brezovsky, J. Hofman, Toxic effects of nine polycyclic aromatic compounds onEnchytraeus crypticus inartificial soilinrelationto their properties, Ecotoxicology and Environmental Safety 74 (2011) 1727–1733.
63. F. Pazos, D. Guijas, A. Valencia, V. De Lorenz, MetaRouter: bioinformatics for bioremediation, Nucleic Acids Research 33 (2005) D588–D592.
64. G. Carbajosa, A. Trigo, A. Valencia, I. Cases, Bionemo: molecular information on biodegradation metabolism, Nucleic Acids Research 37 (2009) D598–D602.

Chapter 8

Analysis of Gas Phase Compounds in Chemical Vapor Deposition of Carbon from Light Hydrocarbons

Koyo Norinaga, Olaf Deutschmann, and
Klaus J. Huttinger

Institut fur Technische Chemie und Polymerchemie, Universita ̈t Karlsruhe, Engesserstr. 20, 76131 Karlsruhe, Germany

ABSTRACT

Product distributions in the pyrolysis of ethylene, acetylene, and propylene are studied to obtain an experimental database for a detailed kinetic modeling of gas phase reactions in chemical vapor deposition of carbon from these light hydrocarbons. Experiments were performed with a vertical flow reactor at 900 °C and pressures from 2 to 15 kPa. Gas phase components were analyzed by both on-line and off-line gas chromatography. More than 40 compounds from hydrogen to coronene were identified and quantitatively determined as a function of the residence time varied up to 1.6 s. Product recoveries were generally more than 90%. Analysis of the kinetics of the conversion of the hydrocarbons resulted in global reaction orders of 1.2 (ethylene), 2.7 (acetylene), and 1.5 (propylene). First order dehydrogenation reactions and third order trimerization reactions leading to benzene are decisive reactions for ethylene and acetylene, respectively. Conversion of propylene should also be based on two simultaneous reactions, a first order dissociation reaction, and second order reactions such as bimolecular reaction of propylene resulting an allyl and a propyl radical. These insights should be useful to develop a reaction mechanism based on elementary reactions in forthcoming studies.

INTRODUCTION

Chemical vapor deposition (CVD) of carbon from light hydrocarbons is mainly used in the production of carbon fiber reinforced carbon (CFC) by infiltrating pyrolytic carbon into carbon fiber performs [1]. The CVD of carbon involves simultaneous processes of complex gas-phase reactions leading to various products including polycyclic aromatic hydrocarbons (PAHs) and soot, and heterogeneous reactions leading to the deposition of pyrolytic carbon on the substrate surface [2] and [3]. Therefore, the overall kinetics of the deposition process is determined by the kinetics of gas-phase and surface reactions and in particular the interaction or competition of gas-phase and surface reactions. A great variety of hydrocarbons and hydrocarbon radicals are formed by gas phase reactions, and any of these species has a potential for chemisorption or physisorption on the growing carbon surface and thus to form pyrolytic carbon. These complications make it difficult to understand the CVD of carbon quantitatively and to develop a precise model. Becker and Hüttinger proposed a simplified model which can describe the kinetics of carbon deposition from C_2 hydrocarbons [4]. A lumping approach was used in which a large number of gas phase species are lumped into three groups, i.e. C_2, C_4, and C_6 hydrocarbons, and the rate constants of the chemical reactions involved are estimated by a numerical fitting. This model was successfully used for simulation of the chemical vapor infiltration (CVI) process [5] and [6]; it can also serve as useful benchmark for detailed kinetic schemes.

It is a challenge for future research to advance from the traditional lumping method and to develop a model based on elementary reactions which describes the complex gas phase chemistry of CVD of carbon at a molecular level. A numerical simulation of CVI based on the detailed chemical kinetics and correlation between kinetics and structures of pyrolytic carbon is necessary to develop and optimize the CFC production process. A great number of studies can be found in literatures for the pyrolysis of hydrocarbons such as ethylene [7], [8], [9], [10], [11], [12], [13], [14], [15], [16], [17], [18], [19], [20], [21], [22], [23], [24],[25], [26] and [27], acetylene [28], [29], [30], [31], [32], [33], [34], [35], [36], [37], [38], [39], [40], [41], [42],[43], [44], [45], [46], [47], [48], [49], [50], [51], [52], [53], [54], [55] and [56], and propylene [57], [58], [59],[60], [61], [62], [63], [64], [65], [66], [67], [68], [69], [70], [71], [72], [73], [74], [75] and [76]. Most of the studies are focused on pyrolysis mechanism at initial stages with shock tube at very short residence times or volume reactor at low temperatures. Although a variety of gas phase products were found in CVD of pyrolytic carbon [77] and [78], experimental results on the hydrocarbon pyrolysis with a wide range of products analysis including large polycyclic aromatic hydrocarbons and pyrolytic carbon are very few. Results obtained at reduced pressures without diluent inert gas, which are employed in CFC production process to gain favorable free gas diffusion, are also insufficient.

Descamps et al. [79] developed a gas phase reaction mechanism (53 species and 205 reactions) in CVD of carbon from propane using elementary reactions reported in the literature. The mechanism was validated by comparing the computations with the experimental data on the gas phase compositions obtained at 2 kPa and two temperatures of 800 and 1000 °C. In situ Fourier transform infrared spectroscopy was employed to determine the concentrations of gas phase components [80] and [81]. Since their analysis was semi-quantitative, their mechanism validation is likely to be insufficient. Detailed and quantitative information on the gas phase components in hydrocarbon pyrolysis at conditions relevant to CVD of carbon is still limited.

Experimental data on the gas phase compositions which are evaluated quantitatively are needed for the development of a more accurate elementary-step like gas phase reaction mechanism. Experiments at well-defined conditions with satisfying material balances are the fundamental requirements. Furthermore, experimental results obtained from various source hydrocarbons are useful to develop a comprehensive mechanism. In this study, the species composition in the gas phase in CVD of carbon from the unsaturated light hydrocarbons ethylene, acetylene, and propylene is analyzed quantitatively. The CVD experiments were performed in a vertical flow reactor. The temperature was 900 °C, pressure was varied from 2 to 15 kPa, and the effective residence time from 0.1 to 1.6 s. Gaseous and condensing products were analyzed using on-line and off-line gas chromatography, respectively.

EXPERIMENTAL

The experimental set-up is given in the Supplementary data. The reactor is identical to that used in a previous study [82]. Total length of the reactor is 440 mm. The deposition space is located at the center of the reactor and formed by a cylindrically shaped alumina ceramic tube, 22 mm i.d. and 40 mm long. A channel structure, made out of cordierite, with 400 channels per square inch is fitted in the alumina ceramic tube, resulting in a relatively high surface area/volume ratio of the deposition space [A/V] of 3.2 mm^{-1}. The inlet and outlet tubes (8 mm i.d.) are connected to the deposition space through conical inlet and outlet nozzles. These narrow tubes increase linear velocity of flowing gas, reducing the predecomposition of the hydrocarbon gas and post reactions beyond the reaction zone. The nozzles are employed to generate a plug-flow [4] which is necessary to obtain reliable kinetic data. Temperature profile for the reactor was measured under an argon flow with a type K thermocouple (Rössel Messtechnik GmbH & Co.) that was moved axially along the reactor length. The measured temperature profile for a set point temperature of 900 °C can be found in the Supplementary data. The axial temperature variation of the deposition space is within ±2 K of the target temperature.

Previous simulation for gas temperature in the reactor applied [83] suggests that the channel structures can effectively homogenize the temperature in the radial direction. Influences of reaction enthalpies on the temperature profiles in axial direction lead to some uncertainties. In ongoing research their effects are studied more accurately by CFD simulations coupled with detailed chemical reaction schemes [84].

The deposition experiments were performed at a temperature of 900 °C, pressures ranging from 2 to 15 kPa and effective residence times of up to 1.6 s. Ethylene, acetylene, and propylene with respective purities higher than 99.4%, 99.6%, and 99.5% were all purchased from Air Liquid Co. Ltd. and used as carbon sources.

The chemically reacting flow leads to density variations along the axial coordinate, which introduces some uncertainties in the estimation of the residence time. The residence time, τ, without taking into account these density changes, here simply derived from

$$\tau = V_R / v_u \tag{1}$$

with V_R = free volume of the deposition space in m³; v_u = the upstream volumetric flow rate of the hydrocarbon gases in m³/s at the deposition temperature, and pressure. Pyrolysis of the hydrocarbons changes the gas density and hence the volumetric flow rate downstream. A corrected flow rate, v_d is therefore used:

$$v_d = v_u * \varepsilon \tag{2}$$

where ε is the volume expansion factor, ε, which is calculated by

$$\varepsilon = v_{out}/v_{in} \tag{3}$$

where v_{out} and v_{in} are the volumetric flow rate at reactor outlet and inlet, respectively, at ambient conditions. The measurement of ε is based on the displacement of a given volume of water per unit of time. Liquid products which would condense at water temperature will cause under-estimations of the exit flow rates. Nevertheless the v_{out} measurement is necessary to calculate the gaseous product yields as well as to establish a perfect material balance. The effective residence time, τ_{eff}, was thus defined by

$$\tau_{eff} = V_R/v_d = \tau/\varepsilon \tag{4}$$

It is impossible to estimate a real or a mean residence time since the residence time distributions are currently not clear. However, it is noted that τ and τ_{eff} represent two extremes of the residence time and the real residence time should lie in between. The CFD simulations coupled with detailed chemical reaction schemes are currently in progress and will help to evaluate the flow rate profile in the reactor accurately.

Gaseous products up to C_4 compounds were analyzed on-line with a Sichromat 3 gas chromatograph (Siemens) equipped with a vacuum dosing system. A Porapak N column (Chrompak) and a thermal conductivity detector were used for separation and peak detection, respectively.

Liquid products larger than benzene were collected in two cold traps set at 195 K, dissolved in a measured amount of acetone and analyzed by a Sichromat 1–4 gas chromatograph (Siemens) equipped with a capillary column (CP-Sil 8 CB LB/MS, Chrompak) and a flame ionization detector. Species were identified by the retention time matching. Details of the analysis of liquid products and a typical chromatogram are given elsewhere [78]. Our previous work of gas phase analysis using methane as a precursor gas [85] indicates that the perfect collection of the products was very difficult with a cold trap after the reactor outlet in CVD experiments at low pressures. The cold trap after the membrane vacuum pump was found to be useful to collect light aromatic hydrocarbons such as benzene and naphthalene and thus enabled to improve the material balances of the analysis. Flexible heaters were maintained at 443 K to avoid condensation and adsorption of products on inner walls. On-line gas phase analysis and off-line analysis of liquid products were performed in separate runs. The relative errors of the product yield generally within ±5% and ±20% for the yields of gaseous and condensing products, respectively.

RESULTS AND DISCUSSION

Compounds Found in Gas Phase

Fig. 1 shows the chemical structures of the compounds found in the gas phase. More than 40 species ranging from hydrogen to coronene could be identified and quantitatively determined as a function of both pressure and residence time. 1-Butene is observed only in the CVD from propylene. Propylene and 1,3-butadiene are not found in the CVD from acetylene. Xylenes and ethylbenzene peaks could not be distinguished. Besides these compounds diacetylene (C_4H_2) and propynylbenzene are observed in the CVD from propane [81]. The aromatic compounds found in the present experiments are identical with those found in our previous CVD experiments at 1100 °C and 5–60 kPa using methane as precursor gas [78]. Propane pyrolysis at 950 °C and 2 kPa [81] as well as ethane pyrolysis at 912 °C and 40 kPa also produce identical aromatic compounds [77]. Polycyclic aromatic hydrocarbons found here are also typical products in the flames of hydrocarbons [86].

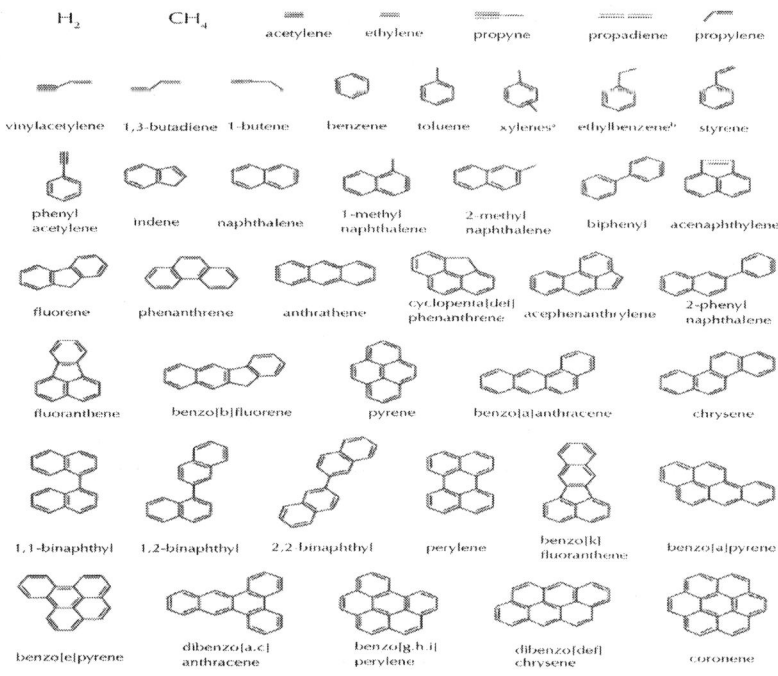

Figure 1: Compounds found in gas phase in CVD of carbon from ethylene, acetylene, and propylene at 900 °C. (a) o-, m-, p-xylenes are not distinguished. (b) Not distinguished from xylenes peaks.

Density Changes of Reacting Flows

Volume expansion factors (ε) and effective residence times (τ_{eff}) at various pressures are presented along with the density change neglected residence times (τ) in Table 1. ε values are larger than 1 with increasing τ for ethylene and propylene experiments, implying that these hydrocarbons principally undergo decomposition reactions leading to an increase in the total numbers of gas phase species. The ε values in propylene experiments are larger than those in ethylene, indicating that the extent of propylene decomposition is more extensive than in the case of ethylene. The ε values in the propylene experiments decrease with increasing pressure, suggesting that decomposition reactions are overlapped by combination reactions being favored at increasing pressures. Unlike these, the ε values of acetylene are less than 1 and decrease with both increasing pressure and τ. Benzene formation by the combination of three acetylene moieties should be responsible for the density increase, making τ_{eff} longer.

Table 1: Prescribed residence time (τ), volume expansion factor (ε), and effective residence time (τ_{eff}) in CVD experiments at 900 °C

Precursor	p, kPa	τ, s	ε, –	τ_{eff}, s
C_2H_4 (ethylene)	2	0.25	1.00	0.25
		0.50	1.02	0.49
		0.75	1.05	0.71
		1.00	1.09	0.92
	4	0.25	1.02	0.24
		0.50	1.03	0.48
		0.75	1.07	0.70
		1.00	1.11	0.90
	8	0.25	1.06	0.24
		0.50	1.05	0.47
		0.75	1.05	0.71
		1.00	1.05	0.95
	15	0.25	1.06	0.24
		0.50	1.08	0.46
		0.75	1.09	0.69
		1.00	1.08	0.93

C$_2$H$_2$ (acetylene)	2	0.25	1.00	0.25
		0.50	1.00	0.50
		0.75	1.00	0.75
		1.00	0.99	1.01
	4	0.25	1.00	0.25
		0.50	0.99	0.51
		0.75	0.94	0.80
		1.00	0.90	1.11
	8	0.25	0.94	0.27
		0.50	0.90	0.56
		0.75	0.88	0.85
		1.00	0.85	1.18
	15	0.25	0.73	0.34
		0.50	0.64	0.78
		0.75	0.61	1.23
		1.00	0.61	1.64
C$_3$H$_6$ (propylene)	2	0.10	1.09	0.09
		0.25	1.29	0.19
		0.50	1.65	0.30
		0.75	1.90	0.39
		1.00	2.17	0.46
	4	0.12	1.07	0.11
		0.25	1.23	0.20
		0.50	1.37	0.36
		0.75	1.61	0.47
		1.00	1.72	0.58
	8	0.12	1.07	0.11
		0.25	1.23	0.20
		0.50	1.42	0.35
		0.75	1.55	0.48
		1.00	1.64	0.61
	15	0.25	1.20	0.21
		0.50	1.37	0.36
		0.75	1.51	0.50
		1.00	1.58	0.63

Material Balances and Product Distributions

Product distributions as well as total carbon yields (material balances) are summarized in Table 2. The yields are calculated based on C_1. The complete sets of the observed species yields are included in theSupplementary data. Gaseous products are distinguished into CH_4, C_2, C_3, and C_4 hydrocarbons, and condensing products are distinguished into benzene and other aromatic hydrocarbons (AHs). Yields of carbon deposited on the substrate were calculated based on the data in our previous study in which the weights of carbon deposited on the substrates were measured [82]. Total carbon yields are generally in the range from 90% to 100%. Carbon deposited outside the substrate and unidentified products account for missing carbon. Total carbon yields slightly higher than 100% may be attributed to experimental errors. In the ethylene experiments the carbon balance is less perfect, especially at lower pressures and longer residence times. The reason may be polymerization products that are formed by thermal polymerization of ethylene [87]. The majority of these polyene compounds would stick to the inner wall of the tubes from which the products were not recovered. Benzene is a major product and comprising much of total condensing products for all precursor hydrocarbons. Yields of benzene and aromatic hydrocarbons increase with increasing pressure and residence time. The yields of deposited carbon are as low as 1–2% for ethylene and propylene. Pressure has a little effect on the carbon yields. On the other hand, the yields of deposited carbon in the CVD from acetylene increase up to 8.9% with an increase in residence time and pressure. The difference in the yields of deposited carbon should result from different carbon to hydrogen ratios of 0.5 (ethylene and propylene) and 1 (acetylene). Hydrogen plays an important role in carbon deposition [88], as it inhibits carbon deposition by blocking active sites, mainly existing at the edges of graphene layers, as a consequence of forming carbon-hydrogen complexes. These results imply that the carbon deposition rate is not simply correlated with the concentration of aromatic hydrocarbons in the gas phase.

Table 2: Product distributions (in %, C_1 base) in CVD experiments at 900 °C

Precursor	p, kPa	τ_{eff}, s	CH_4	C_2	C_3	C_4	Benzene	AHs[a]	PyC[b]	Total
C_2H_4 (ethylene)	2	0.25	0.44	88.9	0.3	3.9	0.4	0.1	1.6	95.7
		0.49	0.77	82.4	0.4	4.3	1.1	0.3	1.9	91.2
		0.71	1.01	76.5	0.5	4.3	1.7	0.4	2.2	86.7
		0.92	1.28	74.2	0.5	4.4	2.2	1.1	2.5	86.1
	4	0.24	0.42	87.5	0.6	6.1	0.7	0.4	1.1	96.8
		0.48	0.88	80.2	0.8	6.4	1.5	1.1	1.3	92.1
		0.70	1.16	74.4	0.8	6.1	2.7	1.9	1.6	88.7
		0.90	1.57	71.5	0.7	5.4	2.9	2.6	1.9	86.6
	8	0.24	0.88	88.3	0.8	5.6	0.9	0.4	0.9	97.7
		0.47	1.75	78.8	1.0	7.0	2.0	1.2	1.1	92.8
		0.71	2.47	72.1	0.9	6.2	3.9	3.1	1.3	90.1
		0.95	3.25	67.7	0.9	5.8	7.0	4.9	1.5	91.1
	15	0.24	1.51	84.3	1.2	7.7	2.5	0.9	0.7	99.0
		0.46	2.77	74.8	1.1	6.6	5.1	2.2	0.9	93.5
		0.69	3.87	68.8	1.1	6.0	8.3	3.7	1.1	92.8
		0.93	4.67	63.1	1.0	5.4	14.2	6.3	1.4	95.9

Analysis of Gas Phase Compounds in Chemical Vapor...

C$_2$H$_2$ (acetylene)									
2	0.25	0.12	94.8	1.1	1.0	0.7	0.7	1.4	99.7
	0.50	0.31	94.3	0.8	0.8	0.7	0.9	1.6	99.5
	0.75	0.38	94.0	0.6	0.8	0.8	1.1	1.7	99.4
	1.01	0.46	92.5	0.6	0.7	0.9	1.5	1.8	98.4
4	0.25	0.18	90.0	1.0	2.5	1.2	2.0	1.9	98.8
	0.51	0.51	84.6	0.8	2.5	1.5	2.6	2.1	94.5
	0.80	0.45	79.4	0.7	2.4	2.8	3.8	2.3	91.9
	1.11	0.58	75.2	0.5	2.0	4.7	5.8	2.6	91.4
8	0.27	0.32	83.0	1.0	4.0	6.8	3.6	3.0	101.7
	0.58	0.51	70.7	0.8	3.5	13.3	5.1	3.7	97.5
	0.85	0.74	62.0	0.6	2.6	18.7	6.9	4.4	95.9
	1.18	1.03	53.1	0.3	2.4	22.1	12.4	5.5	96.8
15	0.34	0.54	59.2	1.0	2.9	17.8	9.1	2.7	93.2
	0.78	1.11	41.7	0.8	1.9	28.0	14.8	3.5	91.7
	1.23	1.65	33.2	0.5	1.4	29.3	16.3	5.3	87.8
	1.64	2.11	27.7	0.3	1.2	32.6	19.6	8.9	92.3

224 Hydrocarbon Chemistry

C$_3$H$_6$ (propylene)										
2	0.09	1.70	4.0	95.9	2.4	n.d.	n.d.	n.d.	n.d.
	0.19	3.72	7.8	78.7	3.3	2.1	0.5	1.3	97.4
	0.30	7.42	15.2	60.1	3.4	2.2	2.6	1.6	92.4
	0.39	10.87	18.1	50.3	4.1	3.2	5.7	1.9	94.2
	0.46	12.79	21.0	42.5	4.1	3.6	4.2	2.2	90.4
4	0.11	2.33	4.9	86.3	2.6	n.d.	n.d.	n.d.	n.d.
	0.20	5.86	11.0	71.4	3.2	4.0	2.7	1.2	99.4
	0.36	9.35	15.9	47.0	2.8	11.0	7.6	1.6	95.1
	0.47	12.63	20.7	35.3	3.3	13.4	9.0	1.9	96.1
	0.58	14.60	22.9	25.5	2.8	18.8	12.4	2.3	99.3
8	0.11	3.26	6.6	75.1	3.7	n.d.	n.d.	n.d.	n.d.
	0.20	6.97	13.1	58.7	4.1	8.2	4.1	0.9	96.1
	0.35	11.24	19.3	32.1	3.4	18.0	9.4	1.2	94.6
	0.48	13.98	22.9	19.6	3.2	22.8	13.2	1.5	97.1
	0.61	15.38	24.7	14.4	3.0	24.8	16.5	1.8	100.6
15	0.21	8.59	15.0	45.5	3.6	10.8	5.5	0.9	89.9
	0.36	10.71	22.0	18.3	3.1	19.9	13.5	1.2	88.7
	0.50	16.66	25.5	9.3	2.8	23.0	16.4	1.5	95.1
	0.63	18.69	27.2	5.9	2.8	29.5	16.6	1.9	102.6

[a] Aromatic hydrocarbons except for benzene.
[b] Pyrolytic carbon deposited on substrate.

Ethylene Pyrolysis

Fig. 2 shows unconverted ethylene as well as product yields in % (based on C_1 for hydrocarbons and H_1 for hydrogen) in CVD of carbon from ethylene at 8 kPa and 900 °C. Ethylene, hydrogen, 1,3-butadiene, methane, acetylene, and benzene are the major compounds. The yield of 1,3-butadiene exhibits a maximum at around 0.4 s whereas the yields of other products increase with increasing τ_{eff}. Possible routes to major products are discussed briefly based on the literature and our recent study of the detailed chemical kinetic modeling of the hydrocarbon pyrolysis [89]. Dehydrogenative decomposition of ethylene is a possible reaction to form hydrogen and acetylene [49]. 1,3-butadiene should be formed principally by combination of ethylene and vinyl radical (C_2H_3) [49], and converts into consecutive products. Reaction of 1,3-butadiene with vinyl radical produce linear C_6 species [90] which further convert into benzene[91] and [92]. Possible routes to explain methane formation are ethyl radical (C_2H_5) decomposition [93] and isomerization of 1,3-butadiene to 1,2-butadiene followed by decomposition into methyl (CH_3) and propargyl (C_3H_3) [94]. Dimerization of propagyls is also known to be an important route to benzene [95].Decomposition of 1-butene which is not observed in this study but detected in other studies [12], [13], [16] and [25] yields methyl and allyl (C_3H_5) and considered as a possible pathway in methane formation.

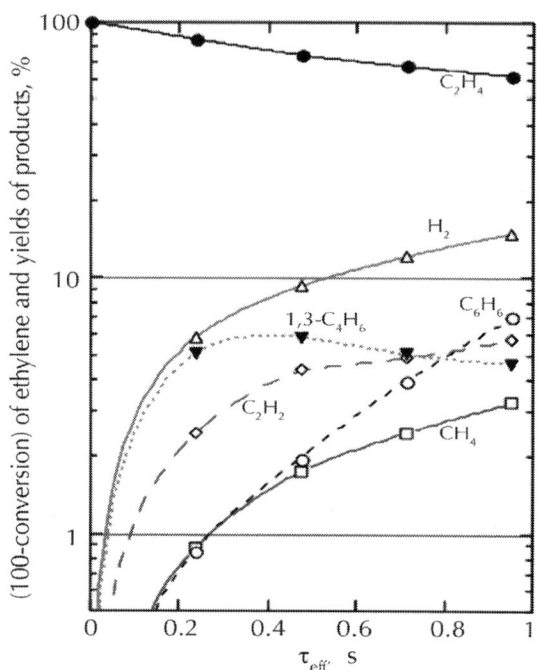

Figure 2: Product yields and (100-conversion) of ethylene vs τ_{eff} in CVD of carbon from ethylene at 900 °C and 8 kPa. Yields are based on C_1 and H_1 for hydrocarbons and hydrogen, respectively.

Acetylene Pyrolysis

Fig. 3 shows uncoverted acetylene as well as product yields in % (based on C_1 for hydrocarbons and H_1 for hydrogen) in CVD of carbon from acetylene at 8 kPa and 900 °C. Benzene, hydrogen, vinylacetylene, naphthalene, ethylene, and methane are major products. The yield of vinylacetylene exhibits a maximum at around 0.2 s whereas the yields of other products increase with increasing τ_{eff}. Hydrogen should be formed by direct formation of pyrolytic carbon from acetylene and formation of PAHs. Dimerization of acetylene yields vinylacetylene [96]. Benzene is mainly formed by combination of acetylene and vinylacetylene [97]. Addition of vinylacetylene to benzene is a possible route in naphthalene formation [98]. In addition to the molecular paths, radical paths involving C_4H_5 and C_4H_3 are also important especially at high temperatures [41] and [99]. Substantial amount of methane should be formed from impurity such as acetone in the acetylene feedstock [47]. Acetone produces methyl radicals that are known to play an important role in branching chain reactions.

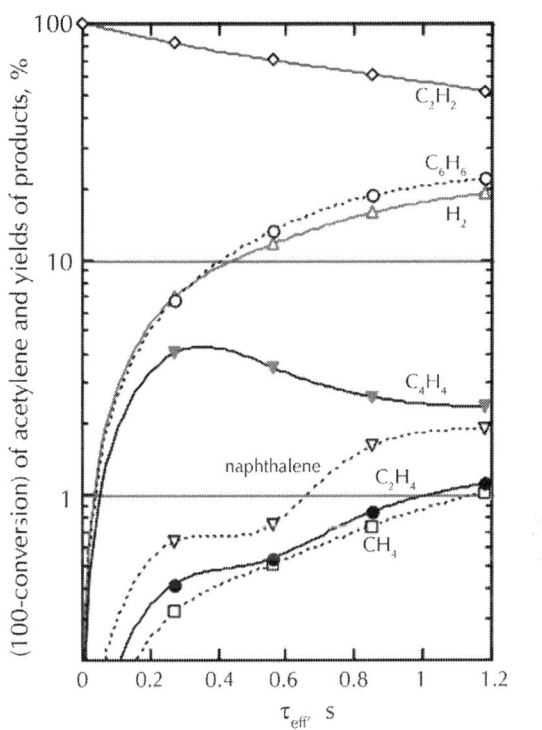

Figure 3: Product yields and (100-conversion) of acetylene vs τ_{eff} in CVD of carbon from acetylene at 900 °C and 8 kPa. Yields are based on C_1 and H_1 for hydrocarbons and hydrogen, respectively.

Propylene Pyrolysis

Fig. 4 shows unconverted propylene as well as product yields in % (based on C_1 for hydrocarbons and H_1 for hydrogen) in CVD of carbon from propylene at 8 kPa and 900 °C. Methane, acetylene, hydrogen, benzene, ethylene, propyne, 1,3-butadiene, and propadiene are major products. The yields of propyne, 1,3-butadiene, and propadiene exhibit maxima at 0.1 ~ 0.2 s whereas the yields of other products increase with increasing τ_{eff}. Methane and ethylene should be formed by α-scission of propylene [73]. β-scission of propylene yields allyl radical (C_3H_5) [72], which is further converted into propadiene [72]. Propyne should be mainly formed by the isomerization of propadiene [100]. These C_3 compounds as well as propargyl radical should play a major role in benzene formation [100].

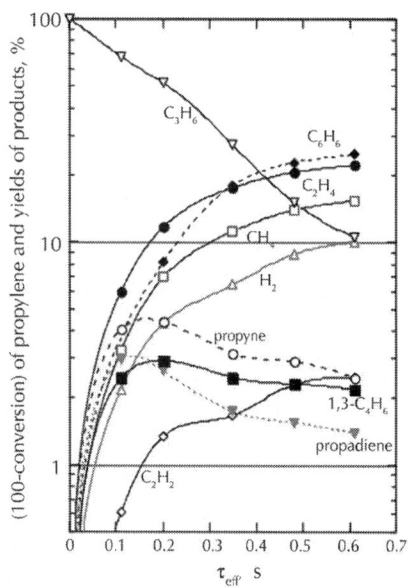

Figure 4: Product yields and (100-conversion) of propylene vs τ_{eff} in CVD of carbon from propylene at 900 °C and 8 kPa. Yields are based on C_1 and H_1 for hydrocarbons and hydrogen, respectively.

Influence of Pressure

The influence of pressure on the conversion of the hydrocarbons is investigated to determine global reaction orders which may provide some information on the reaction mechanism. Fig. 5 shows uncovered amounts in % of ethylene, acetylene, and propylene as a function of the effective residence time at a temperature of 900 °C and various pressures. The conversion of all hydrocarbons increases continuously with increasing τ_{eff}; increasing pressure has a strongly accelerating effect on the conversion of all hydrocarbons, most significantly in the case of acetylene. The accelerating effect of pressure indicates that reactions with an order higher than one are included in the conversion of the hydrocarbons.

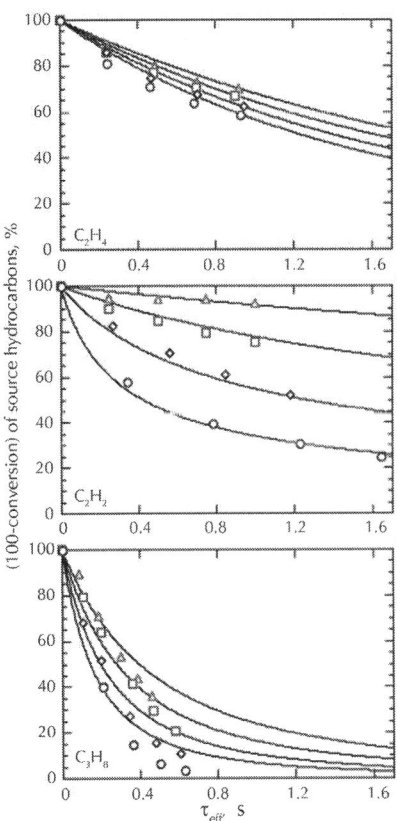

Figure 5: Consumption of source hydrocarbons in CVD of carbon from ethylene(upper), acetylene(middle), and propylene(lower) at 900 °C and pressures of 2 (△), 4 (□), 8 (◊), 15 (○) kPa. The solid lines

were calculated based on a single formal reaction for ethylene (n = 1.2, k = 0.55) and acetylene (n = 2.7, k = 1.5), and propylene (n = 1.5, k = 4.6).

Global Reaction Order and Dominant Reactions

Reaction orders are determined for the formal reaction:

Precursor hydrocarbon → Products

Fractional life method [101] is used to approximate the reaction order. This method is based on

$$t_F = \frac{F^{1-n} - 1}{k(n-1)} C_{i0}^{1-n} \quad (5)$$

where t_F is the time needed for the concentration of reactants to drop to the fractional value of F, k is the global rate constant, n is the reaction order ($\neq 1$), C_{i0} is the initial concentration of species i. Values of t_F are determined at F values of 0.7, 0.8, and 0.6 for ethylene, acetylene, and propylene, respectively. A nnth order rate equation is described by

$$C_i^{1-n} - C_{i0}^{1-n} = k(n-1)\tau_{\text{eff}} \quad (6)$$

where C_i is the concentration of species i at τ_{eff}. Eq. (6) is used to examine whether k is independent of the initial pressure.

Plots based on Eq. (5), that is log t_F vs log C_{i0}, give straight lines of which slopes correspond to (n − 1) (see the Supplementary data). n values were determined to be 1.2 (ethylene), 2.7 (acetylene), and 1.5 (propylene). Previous studies of ethylene pyrolysis showed that the ethylene consumption rate is first order in the ethylene concentration [9] and [11]. The overall reaction order of propylene destruction evaluated here (n = 1.5) also agrees well with previous works by Kallend et al. (n = 1.4) [61] and Kunigi et al. (n = 1.5)

[63]. However the reaction order of acetylene consumption is 2.7 which is higher than second order which seem most assured in the literatures [34] and [38]. Using these n values, plots based on Eq. (6), that is C_i^{1-n} vs $_{eff}$, are made and given in the Supplementary data. The plots can be approximated by straight lines with almost same slopes in the case of ethylene and acetylene, but not of propylene. Averaged k-values resulting from the slopes of the straight lines of these plots are 0.55 and 1.5 for ethylene and acetylene, respectively. The data of acetylene at the lowest pressure of 2 kPa were canceled in averaging k because of their obvious deviations.

For ethylene, n is 1.2 and k is independent of p. The lines drawn in Fig. 5 (upper) are based on Eq. (6) using the values of n = 1.2 and k = 0.55; they show good agreement with the experiments. This result suggests that the overall reaction is dominated by first order decomposition reactions [17] and [20] such as

$C_2H_4 \rightarrow C_2H_3 + H$
$C_2H_4 \rightarrow C_2H_2 + H_2$

A superimposed bimolecular reaction [10],
$C_2H_4 + C_2H_4 \rightarrow C_2H_5 + C_2H_3$

should be responsible for a slightly higher reaction order of n = 1.2.

For acetylene, n is 2.7 and k is almost independent of p except for the data at 2 kPa. The lines in Fig. 5(middle) are also based on Eq. (6) using values of n = 2.7 and k = 1.5. A good agreement with the experimental data is achieved within the range of conditions tested. This suggests a trimerization reaction (n = 3) of acetylene

$3C_2H_2 \rightarrow C_6H_6$

to be the dominating reaction. This result agrees with conclusions of an earlier study [4]. This molecular polymerization represents an overall reaction, in which actually several fragment radical chain steps are included as found by Kieffer et al. [49].

For propylene, plots based on Eq. (5) show that linear approximations are limited to short τ_{eff}. The deviations from linearity become more significant at increasing pressure. Linear approximations were made for short τ_{eff} to determine the k values at all pressures. Using the averaged k of 4.6, n = 1.5, and the Eq. (6) the lines in Fig. 5 (lower) are obtained. A satisfying agreement at short τ_{eff} suggests simultaneously occurring first order and second order reactions in an initial stage of propylene pyrolysis. The first order reactions should include the decomposition reactions such as

$$C_3H_6 \rightarrow C_3H_5 + H$$
$$C_3H_6 \rightarrow C_2H_3 + CH_3$$

The reported rate constants of the above two first order decomposition reactions are 0.18 s^{-1} (β-scission) [72] and 0.03 s^{-1} (α-scission) [73] at 900 °C, implying that the first reaction is clearly privileged. The second order reactions like the bimolecular reaction [65],

$$C_3H_6 + C_3H_6 \rightarrow C_3H_5 + C_3H_7$$

should occur in parallel. Expected differences between calculations and experiments, observed at longer τ_{eff}, are attributed to additional second order reactions. Simon et al. [65] found that the formation rates of 1,3-butadiene, cyclopentadiene and C_6 compounds are second order with respect to propylene initial concentration. This indicates that reactions between propylene and primary reaction products or between primary products leading to C_4, C_5, and C_6 compounds become more significant at longer residence times.

CONCLUSIONS

The product composition in CVD of pyrolytic carbon from ethylene, acetylene, and propylene was analyzed at 900 °C. More than 40 compounds were identified and quantitatively determined as a function of residence time up to 1.6 s and pressures varying from 2 to 15 kPa. Material balances show that more than 90% of carbon could be detected in the experiments, providing a useful experimental database for kinetic modeling studies of gas phase reactions with detailed chemistry [89]. Global kinetic analysis of the conversion of the precursor hydrocarbons provided insight into the decisive reactions occurring in the gas phase, being a first step in the development of a detailed kinetic model based on elementary reactions.

ACKNOWLEDGMENTS

This research was conducted in the Sonderforschungsbereich (SFB) 551 "Carbon from gas-phase: elementary reactions, structures, materials," which is funded by the Deutsche Forschungsgemeinschaft (DFG). The Alexander von Humboldt Foundation is acknowledged for providing a research fellowship to KN.

REFERENCES

1. Golecki I. Industrial carbon chemical vapor infiltration (CVI) processes. In: Delhaes P, editor. World of carbon. Fibers and composites, vol. 2. London and New York: Taylor & Francis; 2003. p. 112–38.
2. Hüttinger KJ. CVD in hot wall reactors—the interaction between homogeneous gas-phase and heterogeneous surface reactions. Chem Vapor Deposition 1998;4(4):151–8.
3. Hüttinger KJ. Fundamentals of chemical vapor deposition in hot wall reactors. In: Delhaes P, editor. World of carbon. Fibers and composites, vol. 2. London and New York: Taylor & Francis; 2003. p. 75–86.
4. Becker A, Hüttinger KJ. Chemistry and kinetics of chemical vapor deposition of pyrocarbon—II. Pyrocarbon deposition from ethylene, acetylene and 1,3-butadiene in the low temperature regime. Carbon 1998;36(3):177–99.
5. Zhang WG, Hüttinger KJ. Chemical vapor infiltration of carbon— revised. Part I: Model simulations. Carbon 2001;39(7):1013–22.

6. Zhang WG, Hüttinger KJ. Simulation studies on chemical vapor infiltration of carbon. Compos Sci Technol 2002;62(15): 1947–55.
7. Skinner GB, Sokoloski EM. Shock tube experiments on the pyrolysis of ethylene. J Phys Chem 1960;64(8):1028–31.
8. Tsang W, Waelbroeck F, Bauer SH. Kinetics of production of C2 during pyrolysis of ethylene. J Phys Chem 1962;66(2):282–7.
9. Gay ID, Kern RD, Kistiako Gb, Niki H. Pyrolysis of ethylene in shock waves. J Chem Phys 1966;45(7):2371–7.
10. Benson SW, Haugen GR. Mechanisms for some high-temperature gas-phase reactions of ethylene acetylene and butadiene. J Phys Chem 1967;71(6):1735–46.
11. Homer JB, Kistiako Gb. Oxidation and pyrolysis of ethylene in shock waves. J Chem Phys 1967;47(12):5290–5.
12. Halstead MP, Quinn CP. Pyrolysis of ethylene. Trans Faraday Soc 1968;64(541P):103–8.
13. Simon M, Back MH. Kinetics of thermal reactions of ethylene. Canad J Chem 1969;47(2):251–5.
14. Roth P, Just T. Measurements on homogeneous thermal decay of ethylene. Phys Chem Chem Phys 1973;77(12):1114–8.
15. Cundall RB, Fussey DE, Harrison AJ, Lampard D. Shock-tube studies of high-temperature pyrolysis of acetylene and ethylene. J Chem Soc Faraday Trans I 1978;74:1403–9.
16. Back MH, Martin R. Thermal-reactions of ethylene at 500 C in the presence and absence of small quantities of oxygen. Int J Chem Kin 1979;11(7):757–74.
17. Tanzawa T, Gardiner WC. Thermal-decomposition of ethylene. Combust Flame 1980;39(3):241–53.
18. Ayranci G, Back MH. Kinetics of the bimolecular initiation process in the thermal-reactions of ethylene. Int J Chem Kin 1981;13(9): 897–911.
19. Koike T, Morinaga K. Shock-tube studies of the acetylene and ethylene pyrolysis by UV absorption. Bull Chem Soc Japan 1981;54(2):530–4.

20. Kiefer JH, Kapsalis SA, Alalami MZ, Budach KA. The very hightemperature pyrolysis of ethylene and the subsequent reactions of product acetylene. Combust Flame 1983;51(1):79–93.
21. Mackenzie AL, Pacey PD, Wimalasena JH. Induction periods in the formation of ethane from the pyrolysis of ethylene. Canad J Chem— Revue Canad Chim 1983;61(9):2033–6.
22. Mackenzie AL, Pacey PD, Wimalasena JH. Radical-addition, decomposition, and isomerization-reactions in the pyrolysis of ethane and ethylene. Canad J Chem—Revue Canad Chim 1984;62(7):1325–8.
23. Jayaweera IS, Pacey PD. The formation of hydrogen in ethylene pyrolysis at 900 K. Int J Chem Kin 1988;20(9):719–29.
24. Dagaut P, Boettner JC, Cathonnet M. Ethylene pyrolysis and oxidation—a kinetic modeling study. Int J Chem Kin 1990;22(6):641–64.
25. Roscoe JM, Jayaweera IS, Mackenzie AL, Pacey PD. The mechanism of ethylene pyrolysis at small conversions. Int J Chem Kin 1996;28(3):181–93.
26. Hidaka Y, Nishimori T, Sato K, Henmi Y, Okuda R, Inami K, et al. Shock-tube and modeling study of ethylene pyrolysis and oxidation. Combust Flame 1999;117(4):755–76.
27. Krestinin AV. Detailed modeling of soot formation in hydrocarbon pyrolysis. Combust Flame 2000;121(3):513–24.
28. Minkoff GJ. Excited states of acetylene and their role in pyrolysis. Canad J Chem—Revue Canad Chim 1958;36(1):131–6.
29. Badger GM, Lewis GE, Napier IM. The formation of aromatic hydrocarbons at high temperatures. 8. The pyrolysis of acetylene. J Chem Soc 1960:2825–7.
30. Asaba T, Hikita T, Yoneda K. Shock tube studies on pyrolysis of lower hydrocarbons. (2) High temperature pyrolysis of acetylene. Kogyo Kagaku Zasshi 1961;64(11):1925.
31. Cullis CF, Nettleton MA, Minkoff GJ. Infra-red spectrometric study of pyrolysis of acetylene. 1. Homogeneous reaction. Trans Faraday Soc 1962;58(474):1117–27.

32. Cullis CF, Nettleton MA. Infra-red spectrometric study of pyrolysis of acetylene. 2. Influence of surface. Trans Faraday Soc 1963;59(482):361–8.
33. Hou KC, Anderson RC. Comparative studies of pyrolysis of acetylene, vinylacetylene, and diacetylene. J Phys Chem 1963;67(8):1579–81.
34. Cullis CF, Franklin NH. Pyrolysis of acetylene at temperatures from 500 to 1000 C. Proc Royal Soc London Ser A—Math Phys Sci 1964;280(138):139–52.
35. Fields EK, Meyerson S. A new mechanism for acetylene pyrolysis to aromatic hydrocarbons. Tetrahedron Lett 1967;8(6):571–5.
36. Back MH. Mechanism of pyrolysis of acetylene. Canad J Chem 1971;49(13):2199–204.
37. Beck WH, Mackie JC. Formation and dissociation of C2 from hightemperature pyrolysis of acetylene. J Chem Soc—Faraday Trans I 1975;71(6):1363–71.
38. Ogura H. Shock-tube study on mechanism of hydrogenation and pyrolysis of acetylene. Bull Chem Soc Japan 1977;50(8):2051–7.
39. Ogura H. Pyrolysis of acetylene behind shock-waves. Bull Chem Soc Japan 1977;50(5):1044–50.
40. Tanzawa T, Gardiner WC. Reaction-mechanism of the homogeneous thermal-decomposition of acetylene. J Phys Chem 1980;84 (3):236–9.
41. Frenklach M, Taki S, Durgaprasad MB, Matula RA. Soot formation in shock-tube pyrolysis of acetylene, allene, and 1, 3- butadiene. Combust Flame 1983;54(1–3):81–101.
42. Duran RP, Amorebieta VT, Colussi AJ. Pyrolysis of acetylene—a thermal source of vinylidene. J Am Chem Soc 1987;109 (10):3154– 5.
43. Wu CH, Singh HJ, Kern RD. Pyrolysis of acetylene behind reflected shock-waves. Int J Chem Kin 1987;19(11):975–96.

44. Chen YQ, Jonas DM, Hamilton CE, Green PG, Kinsey JL, Field RW. Acetylene-isomerization and dissociation. Berichte Der Bunsen-Gesellschaft—Phys Chem Chem Phys 1988; 92(3):329–36.
45. Ghibaudi E, Colussi AJ. Kinetics and thermochemistry of the equilibrium 2(acetylene) = vinylacetylene—direct evidence against a chain mechanism. J Phys Chem 1988; 92(20):5839–42.
46. Kiefer JH, Mitchell KI. Molecular dissociation of vinylacetylene and its implications for acetylene pyrolysis. Energy Fuels 1988;2(4):458–61.
47. Colket MB, Seery DJ, Palmer HB. The pyrolysis of acetylene initiated by acetone. Combust Flame 1989;75(3–4):343–66.
48. Duran RP, Amorebieta VT, Colussi AJ. Radical sensitization of acetylene pyrolysis. Int J Chem Kin 1989;21(10):947–58.
49. Kiefer JH, Vondrasek WA. The mechanism of the homogeneous pyrolysis of acetylene. Int J Chem Kin 1990;22(7):747–86.
50. Benson SW. Radical processes in the pyrolysis of acetylene. Int J Chem Kin 1992;24(3):217–37.
51. Kiefer JH, Sidhu SS, Kern RD, Xie K, Chen H, Harding LB. The homogeneous pyrolysis of acetylene. 2. The High-temperature radical chain mechanism. Combust Sci Tech 1992;82(1–6):101–30.
52. Hidaka Y, Hattori K, Okuno T, Inami K, Abe T, Koike T. Shocktube and modeling study of acetylene pyrolysis and oxidation. Combust Flame 1996;107(4):401–17.
53. Dimitrijevic ST, Paterson S, Pacey PD. Pyrolysis of acetylene during viscous flow at low conversions; influence of acetone. J Anal Appl Pyrol 2000;53(1):107–22.
54. Xu XJ, Pacey PD. An induction period in the pyrolysis of acetylene. Phys Chem Chem Phys 2001;3(14):2836–44.
55. Xu XJ, Pacey PD. Oligomerization and cyclization reactions of acetylene. Phys Chem Chem Phys 2005;7(2):326–33.

56. Vlasov PA, Warnatz J. Detailed kinetic modeling of soot formation in hydrocarbon pyrolysis behind shock waves. Proc Combust Inst 2003;29:2335–41.
57. Amano A, Uchiyama M. Mechanism of pyrolysis of propylene—formation of allene. J Phys Chem 1964;68(5):1133–7.
58. Sakakibara Y. The synthesis of methylacetylene by the pyrolysis of propylene. 1. The effect of pyrolysis conditions on product yields. Bull Chem Soc Japan 1964;37(9):1262–8.
59. Sakakibara Y. The synthesis of methylacetylene by the pyrolysis of propylene. 2. The mechanism of the pyrolysis. Bull Chem Soc Japan 1964;37(9):1268–76.
60. Marshall RM, Purnell JH, Shurlock BC. Initiation of propylene pyrolysis. Canad J Chem 1966;44(22):2778.
61. Kallend AS, Purnell JH, Shurlock BC. Pyrolysis of propylene. Proc Royal Soc London Ser A—Math Phys Sci 1967;300(1460):120–39.
62. Chappell GA, Shaw H. A shock tube study of pyrolysis of propylene. Kinetics of vinyl–methyl bond rupture. J Phys Chem 1968;72(13):4672–5.
63. Kunugi T, Sakai T, Soma K, Sasaki Y. Thermal reaction of propylene—kinetics. Ind Eng Chem Fundamentals 1970;9(3):314–8.
64. Kunugi T, Soma K, Sakai T. Thermal reaction of propylene mechanism. Ind Eng Chem Fundamentals 1970;9(3):319–24.
65. Simon M, Back MH. Kinetics of pyrolysis of propylene. 1. Canad J Chem 1970;48(2):317–25.
66. Simon M, Back MH. Kinetics of pyrolysis of propylene. 2. Canad J Chem 1970;48(21):3313–9.
67. Sims JA, Kershenb Ls, Shroff J. Kinetics of high-temperature pyrolysis of propylene. Ind Eng Chem Process Des Development 1971;10(2):265–71.
68. Thrower PA, Sorg DJ. Pyrolysis of propylene over graphite substrates. Carbon 1973;11(6):672.

69. Burcat A. Cracking of propylene in a shock-tube. Fuel 1975;54(2):87–93.
70. Cundall RB, Fussey DE, Harrison AJ, Lampard D. High-temperature pyrolysis of ethane and propylene. J Chem Soc—Faraday Trans I 1979;75:1390–4.
71. Kiefer JH, Alalami MZ, Budach KA. Shock-tube, laser-Schlieren study of propene pyrolysis at high-temperatures. J Phys Chem 1982;86(5):808–13.
72. Tsang W. Chemical kinetic data-base for combustion chemistry. 5. Propene J Phys Chem Ref Data 1991;20(2):221–73.
73. Hidaka Y, Nakamura T, Tanaka H, Jinno A, Kawano H, Higashihara T. Shock-tube and modeling study of propene pyrolysis. Int J Chem Kin 1992;24(9):761–80.
74. Barbe P, Baronnet F, Martin R, Perrin D. Kinetics and modeling of the thermal reaction of propene at 800 K. Part III. Propene in the presence of small amounts of oxygen. Int J Chem Kin 1998;30(7):503–22.
75. Davis SG, Law CK, Wang H. Propene pyrolysis and oxidation kinetics in a flow reactor and laminar flames. Combust Flame 1999;119(4):375–99.
76. Goos E, Hippler H, Hoyermann K, Jurges B. Reactions of methyl radicals with propene at temperatures between 750 and 1000 K. Faraday Discussions 2001;119:243–53.
77. Glasier GF, Pacey PD. Formation of pyrolytic carbon during the pyrolysis of ethane at high conversions. Carbon 2001;39(1):15–23.
78. Dong GL, Hüttinger KJ. Consideration of reaction mechanisms leading to pyrolytic carbon of different textures. Carbon 2002;40(14):2515–28.
79. Descamps C, Vignoles GL, Feron O, Langlais F, Lavenac J. Correlation between homogeneous propane pyrolysis and pyrocarbon deposition. J Electrochem Soc 2001;148(10):C695–708.

80. Feron O, Langlais F, Naslain R. In situ analysis of gas phase decomposition and kinetic study during carbon deposition from mixtures of carbon tetrachloride and methane. Carbon 1999;37(9):1355–61.
81. Feron O, Langlais F, Naslain R. Analysis of the gas phase by in situ FTIR spectrometry and mass spectrometry during the CVD of pyrocarbon from propane. Chem Vapor Deposition 1999;5(1): 37–47.
82. Norinaga K, Hüttinger KJ. Kinetics of surface reactions in carbon deposition from light hydrocarbons. Carbon 2003;41(8):1509–14.
83. Zhang WGG, Hüttinger KJ. CVD of SiC from methyltrichlorosilane. Part I: Deposition rates. Chem Vapor Deposition 2001;7(4):167–72.
84. Tischer S, Correa C, Deutschmann O. Transient three-dimensional simulations of a catalytic combustion monolith using detailed models for heterogeneous and homogeneous reactions and transport phenomena. Catal Today 2001;69(1–4):57–62.
85. Hu ZJ, Hüttinger KJ. Influence of the surface area/volume ratio on the chemistry of carbon deposition from methane. Carbon 2003;41(8):1501–8.
86. Bockhorn H, Fetting F, Wenz HW. Investigation of the formation of high molecular hydrocarbons and soot in premixed hydrocarbonoxygen flames. Berichte Der Bunsen-Gesellschaft—Phys Chem Chem Phys 1983;87(11):1067–73.
87. Neuschutz D, Zimdahl S, Zimmmermann E. Kinetics of carbon deposition by CVD from ethylene–hydrogen–argon mixtures at 1000–1100 C and 1 bar total pressure. In: Jensen KJ, Cullen GW, editors. Proceedings of the twelfth international conference on CVD. The Electrochem Soc; 1993.
88. Becker A, Hu Z, Hüttinger KJ. A hydrogen inhibition model of carbon deposition from light hydrocarbons. Fuel 2000;79(13): 1573–80.

89. Norinaga K, Deutechmann O. Detailed gas-phase chemistry in CVD of carbon from unsaturated light hydrocarbons. In: Proceedings of fifteenth european conference on chemical vapor deposition. Electrochemical society proceeding, vol. 2005-09, 2005. p. 348–55.
90. Westmoreland PR, Dean AM, Howard JB, Longwell JP. Forming benzene in flames by chemically activated isomerization. J Phys Chem 1989;93(25):8171–80.
91. Weissman M, Benson SW. Mechanism of soot initiation in methane systems. Progress Energy Combust Sci 1989;15(4):273–85.
92. Orchard SW, Thrush BA. Photochemical studies of unimolecular processes. 6. Unimolecular reactions of C6h8 isomers and interpretation of their photolyses. Proc Royal Soc London Ser A—Math Phys Eng Sci 1974;337(1609):257–74.
93. Tabayashi K, Bauer SH. Early stages of pyrolysis and oxidation of methane. Combust Flame 1979;34(1):63–83.
94. Hidaka Y, Higashihara T, Ninomiya N, Masaoka H, Nakamura T, Kawano H. Shock tube and modeling study of 1,3-butadiene pyrolysis. Int J Chem Kin 1996;28(2):137–51.
95. Wu CH, Kern RD. Shock-tube study of allene pyrolysis. J Phys Chem 1987;91(24):6291–6.
96. Duran RP, Amorebieta VT, Colussi AJ. Lack of kinetic hydrogen isotope effect in acetylene pyrolysis. Int J Chem Kin 1989;21(9):847–58.
97. Chanmugathas C, Heicklen J. Pyrolysis of acetylene–vinylacetylene mixtures between 400 C and 500 C. Int J Chem Kin 1986;18(6):701–18.
98. Appel J, Bockhorn H, Frenklach M. Kinetic modeling of soot formation with detailed chemistry and physics: laminar premixed flames of C-2 hydrocarbons. Combust Flame 2000;121(1–2):122–36.
99. Cole JA, Bittner JD, Longwell JP, Howard JB. Formation mechanisms of aromatic-compounds in aliphatic flames. Combust Flame 1984;56(1):51–70.

100. Hidaka Y, Nakamura T, Miyauchi A, Shiraishi T, Kawano H. Thermal-decomposition of propyne and allene in shock-waves. Int J Chem Kin 1989;21(8):643–66.
101. Levenspiel O. Chemical reaction engineering. John Wiley & Sons; 1999.

Chapter 9

Adsorption of Mixed Polycyclic Aromatic Hydrocarbons in Surfactant Solutions by Activated Carbon

Jianfei Liu[a,b], Jiajun Chen[a], Lin Jiang[c], and Xue Yin[a]

[a]State Key Laboratory for Water and Sediment Sciences of Ministry of Education, School of Environment, Beijing Normal University, Beijing 100875, PR China

[b]School of Civil Engineer, Henan Polytechnic University, Jiaozuo 454003, PR China

[c]Beijing Municipal Research Institute of Environmental Protection, Beijing 100037, PR China

ABSTRACT

The adsorption behavior of three polycyclic aromatic hydrocarbons (PAH) in TX100 solution by activated carbon was studied aiming at surfactant recovery. Adsorption experiments were conducted in batch and column models. The adsorption data fitted well with the Langmuir isotherm model, Dubinin–Radushkevich isotherm models and pseudo-second-order kinetics model. Regarding column adsorption, the exhausting time of TX100 was shorter than the breakthrough time of PAH in the fixed AC column. This finding indicates the feasibility of reusing surfactants from soil-washing solutions. The total costs can reduce about $ 0.57 per 10 L washing solution with AC adsorption.

INTRODUCTION

The remediation of hydrophobic organic compounds (HOCs), such as polycyclic aromatic hydrocarbons (PAHs), is a challenging problem [1] and [2]. PAH contamination usually manifests as a complex mixture of compounds resulting from both the combustion of carbon-containing fuels and the accidental or improper disposal of industrial materials [3], [4] and [5]. PAHs are typical hydrophobic organic molecules that contain one or more benzene rings arranged in various configurations. They are found in high concentrations especially in many industrial sites and urban soils [6]. PAHs are relatively stable and less bioavailable for microbial degradation than many other organic compounds [7] and [8]. Therefore, PAHs are also classified as persistent organic pollutants because they tend to remain in the environment for a long time.

Surfactant-enhanced aquifer remediation (SEAR) has been proposed as a promising remediation technology because of its relatively high removal ratio and cost effectiveness [9], [10] and [11]. SEAR is implemented by injecting dilute aqueous surfactant solutions into contaminated soil. Surfactant solutions can lower the interfacial tension between organic and aqueous phases, as well as increase the apparent solubility of HOCs via micelle solubilization above its critical micelle concentration (CMC) [12], [13] and [14]. The most serious impediment of the full-scale application of the technology is the cost of the surfactant due to surfactant cost can increase operational expenses by up to 50% [15]. Separation of organic contaminants from surfactant solutions is necessary to reuse surfactant solutions and to lessen the demand on waste disposal. Many processes can be employed to remove HOCs from contaminated surfactant solutions based on different principles. Separation processes through adsorption have the advantages of low operational and maintenance cost, low pollution, and few equipment investments. Given the high production cost of activated carbon (AC), AC should be regenerated or reused. Adsorption technology can be utilized as a surfactant recovery method in surfactant-enhanced remediation.

However, studies on the reuse of surfactant solutions by AC are rare. It was reported that adsorption is not practical for the treatment of high concentrations of waste for surfactant reuse by AC [16]. However, Ann proposed that adsorption by AC is an ideal alternative for recovering surfactants because the partitioning coefficient of phenanthrene (PHE) is higher than that of nonionic surfactants [17], [18] and [19]. Wan also verified the capability of activated carbon on selective removal of contaminant while the surfactant was recovered and reused [20].

Considering that contaminants naturally exist as a mixture, the adsorption characteristics of mixed contaminants may be different from those of single contaminants in surfactant solutions. The present study aims to investigate the performance of AC in the selective adsorption of HOCs from nonionic surfactants and to address the problems associated with multicomponent adsorption from surfactant solutions. A mixture of three PAHs often found together in soil-washing solutions was studied, namely, PHE, fluoranthene (FLA), and benzo (a) anthracene (BaA). TritonX-100 (TX100) was used as a nonionic surfactant because target pollutants can be removed at a larger amount and fewer losses are incurred by sorption onto the soil [21], [22], [23], [24] and [25]. Adsorption experiments were carried out both in batch and fixed-bed column systems. The parameters of the adsorption isotherm and kinetics models were determined. This work was supposed to provide further information on the applicability of surfactant reuse by AC adsorption.

MATERIALS AND METHODS

Materials

The nonionic surfactant TX100 with purity >98% was purchased from J&K Scientific Ltd. TX100 is a nonionic surfactant with a molecular weight of 625 g/mol and CMC of 150 mg/L. Three polycyclic aromatic hydrocarbons, PHE, FLA, and BaA (Tkkyo chemical industry Co. Ltd., purity >97%) were employed in the studies. Selected physical and chemical properties of these PAHs are provided in Table 1. All the above chemicals were used without further purification.

Table 1: Selected properties of three PAHs

PAHs	MS [26]	MW (g/mol)[26]	MD (Å × Å × Å)[26]	Cs 25 °C (µg/L) [27]	Log K_{ow}[28]
Phenanthrene PHE		178.2	11.7 × 8.0 × 3.4	1290	4.57
Fluoranthene FLA		202.26	11.4 × 9.5 × 3.8	200–260	5.22
Benzo(a)anthracene BaA		228.3	13.9 × 3.7 × 3.9	10	5.91

MS: molecular structure; MW: molecular weight; MD: molecular dimension; Cs: water solubility; K_{ow}: octanol–water partition coefficient.

Commercial AC supplied by Sinoharm Chemical Reagent Co., Ltd., China, was used as the adsorbent. The AC was boiled in deionized water for 1 h, purged with deionized water for 3–5 times, then dried under 105 °C over night and stored in a dryer before use. The pore structure characteristics of AC were determined by nitrogen adsorption at 77 K using ASAP instrument. The Brunauer–Emmett–Teller (BET) surface area and pore size distribution were obtained by BET equation and density function theory method. Carbon (C), hydrogen (H), and nitrogen (N) contents of the adsorbents were determined by an element analyzer and oxygen (O) content was calculated by difference.

Solubilization Experiments

For each batch test, excess amount (about four times of the apparent solubility) of an individual PAH was added to each glass tubes containing a series of 25 mL TX100 surfactant solutions with various concentrations to ensure maximum solubility. These samples were then equilibrated for a period of 48 h on a reciprocating shaker at approximately 200 rpm maintained at a temperature of 25 ± 0.5 °C. After shaking, the vials were allowed to settle for at least 15 h then subsequently centrifuged at 4000 rpm for 20 min to separate the undissolved PAHs. The solubilization experiments were duplicated and the average values were calculated.

Batch Adsorption Experiment

A stock solution was prepared by dissolving individual or mixed PAHs in TX100 solution. The concentration of PHE, FLA, BaA, and TX100 is 120, 80, 12, and 5000 mg/L respectively in stock solutions. This was diluted to obtain the required concentrations for further use. The activated carbons (1 g/L) were added to 150 mL glass flasks and then filled with 100 mL solution containing both TX100 and individual or mixed PAHs. The adsorption equilibrium isotherms for individual or mixed PAHs were carried out on a reciprocating shaker at approximately 200 rpm maintained at a temperature of

25 ± 0.5 °C for 24 h. Blank samples (without carbon) were also used to determine the value of PAH or TX100 volatilization and adsorption on the surface of the glass flask. The blank recoveries ranged from 94% to 98% and the data were adjusted for these recoveries. In adsorption kinetics experiment, the aqueous samples were taken at 10, 20, 30, 40, 60, 120, 180, 240, 480, 600, 900, 1080, and 1440 min. After shaking, the suspension was centrifuged at 5000 rpm for 10 min.

Fixed Bed Adsorption Experiment

The fixed bed experiment was conducted using organic glass column of 2.5 cm inner diameter and 20 cm height. The column was flushed with de-ionized water for 1 h and dried at 80 °C for 24 h, then filled with activated carbon. Solution was prepared by dissolving PAHs at a known mass in TX100 solution. The various parameters like bed depths (4.7, 10.8, and 14.7 cm), flow rates (10, 20, and 30 mL/min), influent TX100 concentration (3, 4, and 5 g/L), and influent PAH concentration (PHE concentration is 60, 90, and 120 mg/L, respectively) were taken in the fixed bed experiment. The solutions at the outlet of the column were collected at every 15 min.

Regeneration of the Saturated Activated Carbon

After adsorption, AC was dried at 105 °C for 6 h and regenerated in a quartz reactor by 2500 MHz microwave irradiation at 800 W for 6 min. The experimental setup have been described elsewhere [29]. The regeneration activated carbon was reused for further cycles of adsorption in a batch study.

Analytical Methods

After the filtration with 0.2 μm PTFE filter, PAH and TX100 were analyzed by using high-performance liquid chromatography (HPLC,

Dionex U3000) with an ultraviolet detector and an Agilent PAH column (250 × 4.6 mm) packed with 5 μm particles. And HPLC conditions were the following: Flow rate of 1.0 mL/min, injection volume of 20 μL, UV wavelength of 230 nm, and isocratic mobile phase flow with acetonitrile:water = 85:15.

RESULTS AND DISCUSSION

Physical Characterization of AC

Selected physical characteristics of AC are listed in Table 2. The longitudinal length of BaA (the longest length of three PAHs) is 1.4 nm. The monomer and the aggregation size of TX100 are 2.7 nm [30] and 11.6 nm [31], respectively. It is observed that most of the micropores of the AC would be within 0.3 and 2.0 nm from Fig. 1. TX100 in monomer form cannot enter the micropores of AC, not to mention the TX100 in micelle form. Therefore, the AC would be appropriate material for selective adsorption of PAHs dissolved in surfactant solution.

Table 2: The specific properties of activated carbon

BET surface area (m^2/g)	718.2
Total pore volume (cm^3/g)	0.845
Average particle diameter (mm)	0.78
Pore size (nm)	0.2–2.0
Ash (wt. %)	1.12
C (wt. %)	96.47
H (wt. %)	0.48
N (wt. %)	0.36
O (wt. %)	1.57

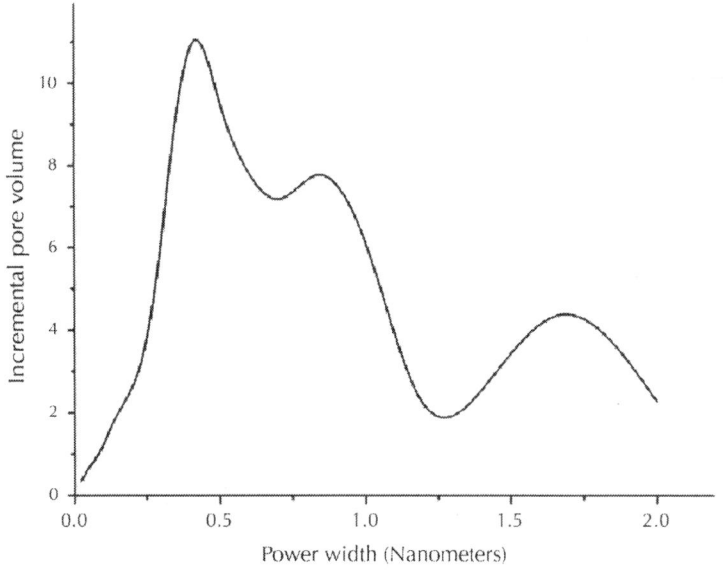

Figure 1: Pore size distributions of activated carbon.

Solubilization of Individual Compounds

The solubility enhancement of three PAHs as a function of surfactant concentration above the CMC is illustrated in Fig. 2. PAHs solubilized in the micelles increased linearly with the increase of surfactant concentration [21]. The slopes of the lines, which represent weight solubilization ratio (WSR) were calculated using least-square linear regression. As for TX100, WSR for PHE, FLA and BaA is 0.033, 0.026, and 0.005, respectively. The solubility of PHE, FLA and BaA in micellar solutions of TX100 (5 g/L) was 0.145, 0.108 and 0.022 g/L, respectively. The solubility of PAH can be increased or decreased by interactions between components in mixed system [32], [33] and [34]. Therefore, the maximum concentrations of each PAH are below their individual solubility in the mixed system. In the following experiment, the maximum concentration of PHE, FLA and BaA is 0.12, 0.08, and 0.012 g/L in the mixed PAH system in the surfactant solutions, respectively.

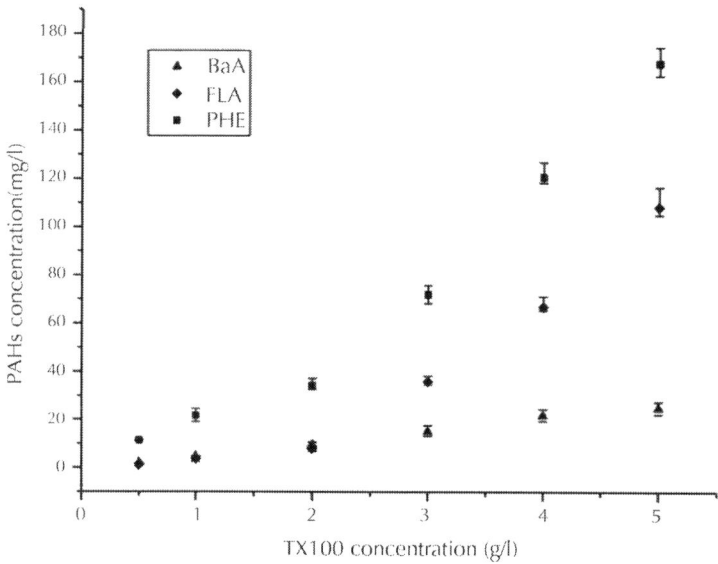

Figure 2: Solubilization of PAHs by the nonionic surfactant TX100.

Adsorption Isotherms of PAHs in the Presence of Surfactants

The equilibrium adsorption isotherm is important in the design of adsorption systems. Fig. 3a–c illustrates the adsorption isotherms of single or ternary PAHs in the presence of surfactants. The symbol of single or ternary PAHs represents one or three mixed PAH in the surfactant solution system. Two commonly used empirical adsorption models, Langmuir [35] and Freundlich [36], were employed to facilitate the estimation of adsorption capacity. Langmuir isotherm model describes monolayer adsorption onto homogeneous surface with no interaction between adjacent adsorbed molecules. Freundlich isotherm model describes adsorption process on heterogeneous surfaces and is suitable to describe adsorption in a narrow range of solute concentration. The linearized forms of two models can be represented in Eqs. (1) and (2) as follows:

$$\frac{1}{q_e} = \frac{1}{q_{max}b}\frac{1}{C_e} + \frac{1}{q_{max}} \qquad (1)$$

$$\log q_e = \frac{1}{n}\log C_e + \log K_F \qquad (2)$$

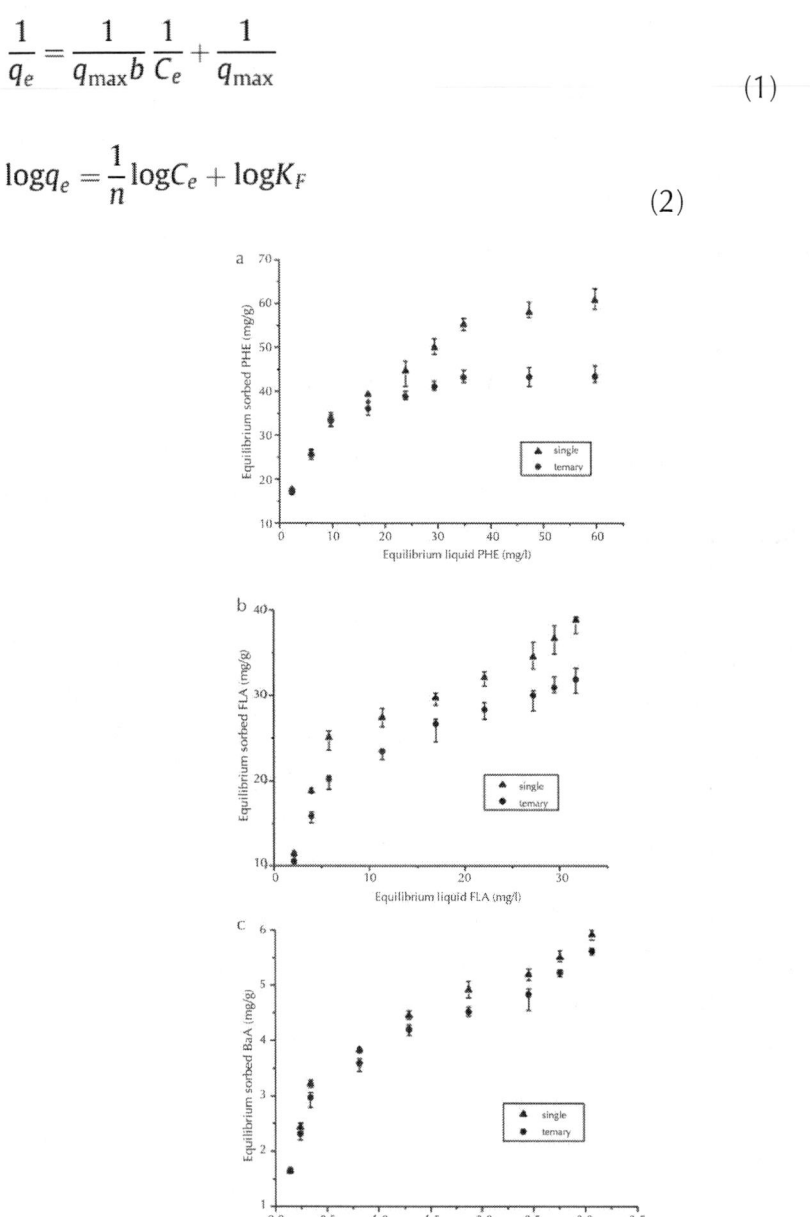

Figure 3: (a) Adsorption isotherms of PHE in TX100 solution. (b) Adsorption isotherms of FLA in TX100 solution. (c) Adsorption isotherms of BaA in TX100 solution.

Eqs. (1) and (2) represent Langmuir and Freundlich model respectively, where q_e is the amount adsorbed per unit mass of adsorbent at equilibrium concentration (mg/g); C_e is the equilibrium concentration of the adsorbate (mg/L); q_{max} is the maximum adsorption capacity (mg/g); b is the adsorption equilibrium constant, characteristic of the affinity between the adsorbent and adsorbate; KF is a Freundlich constant representing the adsorption capacity (g/L); n refers to the adsorption capacity.

The experimental isotherm data were fitted to these equations by applying linear regression analysis. The coefficients of these two isotherm models have been shown in Table 3. The data provide information on the maximum amount of AC required to adsorb the amount of PAHs under specific system conditions. Correlation coefficients are also calculated by fitting the experimental adsorption equilibrium data for two models.

q_{max}, the most important parameter governing the ultimate capacity of AC for adsorbing PAHs in surfactant solutions, varied significantly among the different types of PAH. The trend in adsorption capacity was the opposite of the K_{ow} of PAH. This result agrees with the conclusion of another study [37]. The q_{max} of PHE was 58.843 mg/g, which was several times higher than that of FLA or BaA. With one adsorbent used, the difference in adsorbed amount was dependent on the concentration and structure of PAHs. Jovan lemic [38] found that under the same single PAH concentration, the adsorption is higher for compounds with larger K_{ow}. Therefore, the difference in adsorbed amount was largely attributed to the difference in their concentration, i.e., the amount of PAHs available for a unit amount of AC.

Table 3: Langmuir, Freundlich and Dubinin–Radushkevich adsorption isotherm constants

		Langmuir constant			Freundlich constant			Dubinin–Radushkevich		
		q_{max}	b	r^2	K_F	n	r^2	q_m	E	r^2
Single	PHE	58.84	0.0944	0.9823	0.193	2.5883	0.9612	7.86	1.3853	0.9935
	FLA	39.98	0.1265	0.9773	0.142	2.5828	0.9325	5.62	1.3881	0.9904
	BaA	6.11	0.1288	0.9941	0.065	2.2573	0.9808	2.51	1.8665	0.9859
Mixed	PHE	44.09	0.0675	0.9896	0.120	4.5560	0.9629	8.24	1.2972	0.9951
	FLA	36.36	0.1575	0.9996	0.083	2.6199	0.9605	4.59	1.2972	0.9859
	BaA	6.00	0.0687	0.9875	0.055	2.5644	0.9575	1.47	1.3367	0.9971

In measuring adsorption in mixed PAH, the above equations were applied to each PAH, ignoring the possible interference among PAHs. The possible reduction in adsorption in mixed solutions was evaluated by comparing the maximum adsorption of the single PAH system to that of the mixed PAH system in surfactant solutions. A reduction in single PAH adsorption capacity in mixed PAHs is common in surfactant solutions, although the extent of reduction varies with the type of PAH. The reduction can be attributed to the interaction and competitive adsorption between PAHs. The reduction rates of PHE, FLA, and BaA were 25%, 9%, and 1.8%, respectively. The reduced adsorption value of PAH in the mixed system decreased as the molecular size of PAH increased. As shown in Table 3, the reduced adsorption value of each PAH in the mixed system increased as the initial concentration of PAH increased. The combined adsorbed mass of the three PAHs in the mixed system was 86.45 mg/g (44.09 + 36.36 + 6.00 respectively), which was higher than that in the single PAH system. An increase in total solid phase adsorbent concentration in the mixed system was also found by previous researchers in the adsorption of dye from solution by AC [39].

In adsorption studies, the adsorption mechanism is necessary to be determined. The Dubinin–Radushkevich model was used to determine the characteristic porosity and the mean free energy of adsorption [40]. Different from the Langmuir or Freundlich isotherm model, the Dubinin–Radushkevich isotherm model does not assume either the homogeneous or heterogeneous surface. The linear presentation of this model is expressed by

$$\ln q_e = \ln q_m - \beta \varepsilon^2 \qquad (3)$$

$$E = \frac{1}{\sqrt{2\beta}} \qquad (4)$$

Where q_m is the theoretical saturation sorption capacity based on the Dubinin–Radushkevich isotherm (mg/g); β is related to mean adsorption energy (kJ/mol); ε is equal to $RT\ln(1+1/C_e)$, R is gas constant (kJ/mol K), T is temperature (K) and E is the mean adsorption energy (kJ/mol).

E is the change in free energy when one mole in the solution is transferred to the surface of the membrane from infinity. As presented in Table 3, the E value of PAH adsorption in TX100 solutions by AC ranged from 1.29 kJ/mol to 1.86 kJ/mol. Given that all E values were below 8, the adsorption process was considered to be a physical adsorption [41].

Adsorption Isotherms of the Surfactant

The symbol of TX100 or TX100 + PAH represents individual TX100 or TX100 mixed with PAH in Fig. 4, respectively. As seen in Fig. 4, the TX100 adsorption process in the absence and presence of PAH on AC was rapid in the first step and then nearly reached a plateau in the second step, which is also observed in other reports [42] and [43]. The fitting parameters of Langmuir and Freundlich model for the all cases of TX100 in the absence and presence of PAH are summarized in Table 4. It was reported that adsorption isotherms of nonionic surfactants on AC had a monolayer adsorption followed by surface aggregation above CMC [44], [45] and [46]. However, the Langmuir model fitted the isotherm data better than the Freundlich model in terms of higher regression coefficients. This finding can be attributed to the non-formation of aggregate and semi-aggregate of TX100 inside the micropores because aggregate sizes were larger than pore sizes. Therefore, the adsorption process can be assumed as the formation of an adsorbate monolayer on the outer surface of the adsorbent, with no further adsorption thereafter.

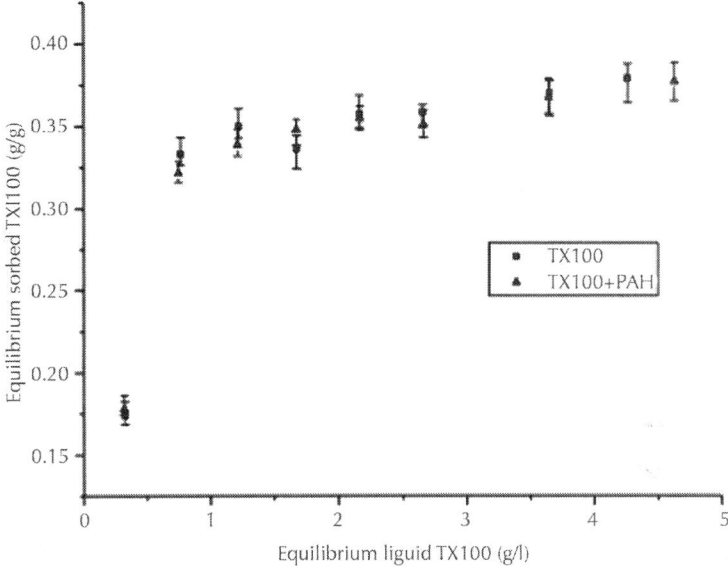

Figure 4: Adsorption isotherms of TX100.

Table 4: Langmuir and Freundlich of TX100 adsorption isotherm constants

	Langmuir constant			Freundlich constant		
	q_{max}	b	r^2	KF	n	r^2
TX100	436.7	0.9587	0.9467	0.291	4.2280	0.8036
TX100 + PHE	437.3	1.8546	0.9087	0.294	2.9250	0.8151
TX100 + FLA	434.3	1.4895	0.9809	0.295	3.9397	0.7828
TX100 + BaA	438.5	1.0255	0.8725	0.300	3.8568	0.6936
TX100 + Tern	437.9	1.0813	0.9763	0.295	2.7386	0.8394

The adsorption isotherm of TX100 in the presence of PAHs is also shown in Fig. 4. Considering that the adsorption isotherm of TX100 with any PAH is similar and that the difference among PAHs is small, only one isotherm of TX100 with PAH was obtained as an example. The presence of PAHs did not affect the isotherm shape of TX100, and q_{max} of the adsorbed TX100 had no significant change in all cases. This finding indicates that the adsorption value of TX100 was not affected by co-existing PAHs, which is consistent with previous research on the adsorption of another solute in surfactant solutions [19] and [47]. One possible reason for this result is that the adsorption capacity of AC for TX100 reached its maximum value and showed no significant change regardless of the presence of single or mixed PAHs. Considering that PAHs often exist as a mixture in wastes, surfactant recovery is beneficent in practice.

Adsorption Kinetics

The evolution of the concentration of PAH as a function of time, plotted as C/C_0 against time is shown in Fig. 5 for TX100 and each PAH in mixed system by AC adsorption. As seen in the Fig. 5, the PAH and TX100 adsorption process on AC can be rapid in the first 30 min, then continued with a slower rate until 24 h of the experiment nearly reached a plateau and the adsorption of PAH was slower than that of surfactant on activated carbon.

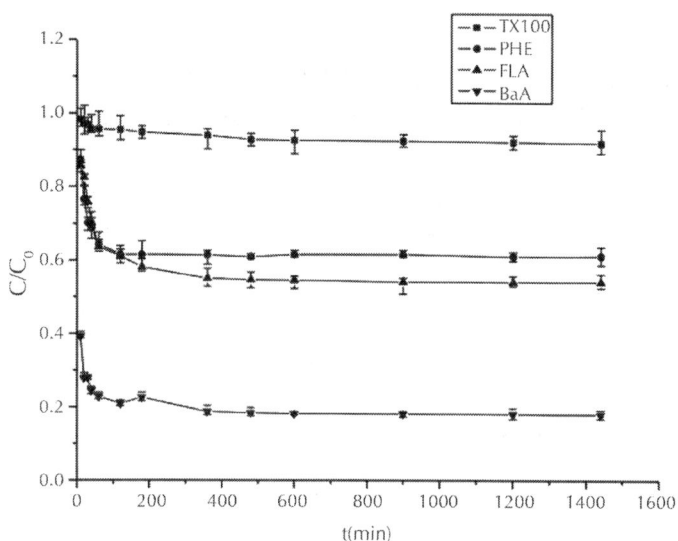

Figure 5: Time schedule of adsorption of PAH and TX100 on activated carbon.

The modeling of the adsorption kinetics may be described by two common models, namely, the pseudo-first-order rate model and a pseudo-second-order model. The pseudo-first-order kinetics equation has been widely used for the adsorption of an adsorbate from an aqueous.

The pseudo-first-order kinetics model [48] and the integral form of this model were expressed by the following equation.

$$\log \frac{q_e}{q_e - q_t} = k_1 t \tag{5}$$

Where q_e and q_t are the amounts of adsorbate adsorbed on the adsorbent at equilibrium and at any time t (mg/g); k_1 is the rate constant of the first-order adsorption (min^{-1}). The straight-line plots of $\log (q_e - q_t)$ against t were used to determine rate constant k_1.

The pseudo-second-order kinetics model [49] and the integral form of this model were expressed by the following equation.

$$\frac{t}{q_t} = \frac{1}{k_2 q_e^2} + \frac{t}{q_e} \tag{6}$$

Where k_2 is the rate constant of the pseudo-second-order model (g/mg min). Values of k_2 and q_e were calculated from intercept and the slope of the linear plots of t/q_t against t.

The constants of the two kinetics models were shown in Table 5. It can be seen clearly that the data agrees well with the pseudo-second-order model from R^2 and S.D. % values and the calculated q_e values agreed with the experimental q_e values. This suggests that the adsorption of PAH on AC follows second-order kinetics. Similar phenomena have been observed in the adsorption of a drug on in the presence of surfactants [46]. It should be noted that a higher correlation coefficient in the first-order kinetics could be obtained if the fitting analyses was made in shorter contact time. But in the whole, the data did not agree with the pseudo-first-order kinetics model.

Table 5: The adsorption kinetic models of PAH in TX100 solution

	C_0	q_{exp}	Pseudo-first-order				Pseudo-second-order			
			k_1	$q_{e,cal}$	R^2	S.D. %	$k_2 \times 10^{-3}$	$q_{e,cal}$	R^2	S.D. %
PHE	120	46.309	0.018	42.266	0.904	6.89	0.084	46.957	0.999	1.27
FLA	80	35.976	0.013	38.103	0.918	13.72	0.427	37.556	0.998	1.22
BaA	12	9.748	0.023	8.447	0.864	5.46	4.901	9.881	0.998	1.49

q_{exp} is the amount of PAH adsorbed on the adsorbent at equilibrium which obtained from experiment, $q_{e,cal}$ is the amount of PAH adsorbed on the adsorbent at equilibrium, S.D. is the normalized standard deviation.

It can also be seen that the second-order adsorption k_2 increases with the decrease of the initial concentration of PAH. And the results are consistent with the other study [50] in which it was found that k_2 is linearly related to K_{ow} for the correlation coefficient between $\log K_2$ and $\log K_{ow}$ is 0.9904.

Fixed-bed Adsorption

Fixed bed adsorption experiments are necessary for designing of field scale adsorption system. The breakthrough curve for adsorption of PAHs and TX100 in the effluent of the fixed bed is shown in Fig. 6. The curve was plotted in accordance with C/C_0 against t. To reuse surfactant, we hope that the AC can remove approximately 90% of PAHs mass meanwhile recover approximately 90% of surfactant mass.

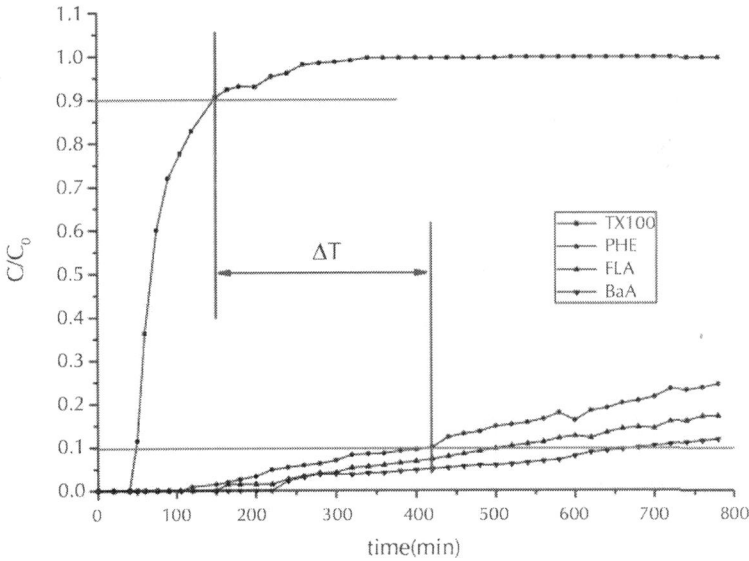

Figure 6: Breakthrough curves for PAHs in TX100 solution by fixed activated carbon column adsorption (initial concentration of TX100, PHE, FLA and BaA is 5, 0.12, 0.08 and 0.012 g/L, respectively, flow rate is 20 mL/min).

As seen from Fig. 6, TX100 appeared first, followed by three PAH in the condition of flow rate is 20 mL/min with 20 g AC (14.7 cm), and initial concentration of TX100, PHE, FLA and BaA is 5, 0.12, 0.08 and 0.012 g/L, respectively. Similar phenomenon was observed in the adsorption of chlorinated solvents in the presence of surfactants [51]. For TX100, the adsorption capacity is 0.28 g/g. The breakthrough time (corresponding to $C/C_0 = 0.1$) and exhausting time (corresponding to $C/C_0 = 0.9$) of TX100 is 66 min and 150 min, respectively. For PAHs, all three PAH breakthrough curves exhibited the same trend. However, the breakthrough time increased with decreasing initial concentration. The breakthrough time of PHE, FLA, and BaA were approximately 420, 520, and 700 min, respectively. Within the period of 150–420 min (T in the Fig. 6), the effluent concentration of TX100 retained over 90% of the initial concentration, whereas the concentrations of the PAHs were below 10% of their initial concentration. We call T as an effective surfactant recovery time. This behavior of various solutes at different times indicates that high surfactant recovery can be achieved by a fixed AC column. Although a few PAHs still remained in the solution, the solution can be reused for another soil washing with the addition of a fresh surfactant solution because of its lower PAH concentration. After 420 min, the concentrations of the PHEs exceeded 10% of the initial concentration. Hence, the solution cannot be reused for environmental protection. This indicates that surfactant can be reused for a long time from the effluent surfactant concentration retain enough much concentration to effluent concentration of one contaminant over a regulatory level in a field washing process. After 700 min, the concentrations of the three PAHs were over 10% of their initial concentration. Although the contaminant concentration did not reach the exhaustion point, the fixed-bed adsorption experiment was terminated.

When initial concentration of TX100, PHE, FLA and BaA is 5, 0.12, 0.08 and 0.012 g/L, respectively, and flow rate is 20 mL/min, 1 g activated carbon treats around 270 mL soil washing solution. It is not satisfying for its high cost, however, we can control the operate condition to reduce the mass of activated carbon depend on the contamination level and the treatment goal. Thus, various parameters were investigated and the result were listed in the Table 6 Since the breakthrough time of PHE are shorter than that of FLA and BaA and the effective surfactant recovery time depended on the breakthrough time of PHE, the breakthrough time of FLA and BaA didn't appeared in the Table 6. As shown in the Table 6, the breakthrough time of both TX100 and PAH increased with the increase of bed height and decrease of flow rate and influent concentration. In practical engineering, different pollutants in the solution have different concentrations and the pollutant concentrations are generally lower than our experimental concentration. The change in contaminant concentration affects the saturation rate of activated carbon and breakthrough time because a lower contaminant concentration caused a lower transport attributed to a decreased diffusion coefficient or mass transfer coefficient. The lower contaminant concentration could lead to longer effective surfactant recovery time, which indicates activated carbon could treat more solutions. In our optimization condition, when initial concentration of TX100, PHE, FLA, and BaA is 5, 0.06, 0.04, and 0.006 g/L, respectively, and flow rate is 10 mL/min, 10 L soil washing solution can be treated by 7.94 g activated carbon.

Table 6: The breakthrough time of PHE and the exhausting time of TX100

$C_{0,TX100}$ (g/L)	$C_{0,PHE}$ (mg/L)	Rate (mL/min)	Depth (cm)	$t_{TX100,0.9}$ (min)	$t_{PHE,0.1}$ (min)	ΔT (min)
5	120	20	7	86	61	−25
5	120	20	10.8	105	165	60
5	120	20	14.7	150	420	270
5	120	10	14.7	302	689	387
5	120	30	14.7	106	196	90
5	90	20	14.7	160	505	345
5	60	20	14.7	162	588	426
4	60	20	14.7	190	606	416
3	60	20	14.7	214	610	396

The initial concentration of the surfactant was significantly higher than that of the PAHs, and the surfactant was the first to reach the breakthrough time, which can elicit negative effects on PAH adsorption. However, an effective surfactant recovery can be achieved by a fixed AC column. This finding may be attributed to the fact that surfactant aggregation in the AC surface can increase surfactant concentration, thereby allowing a higher degree of PAH solubilization. An increase in adsorption capacity due to surface solubilization was also found in previous batch experiments by AC [19], [44] and [47]. This result suggests that fixed AC column adsorption can be used as an alternative method for surfactant-enhanced remediation.

Cost Appraisal

Based on the surfactant and AC efficiency, used surfactant efficiency, the costs of surfactant and AC, AC recovery cost, this section presents a rough economic analysis for surfactant reuse at the experimental scale.

Surfactants are used as additives to enhance the efficiency of soil washing for PAH-contaminated soils treatment. It was discovered that surfactant had the highest washing efficiency between 83% and 90% [52], [53] and [54] and the surfactant remained in liquid approximately 90% after washing. In the AC adsorption experiment, 90% of surfactant remained meanwhile 90% of PAH was removed in fixed bed experiment. Thus, surfactant could be reused after separation of AC from the solution.

The efficiency of the recovered surfactant is an important factor affecting recovery of surfactant in soil washing solutions with activated carbon. In term of overall cost efficiency, a small number of surfactant reuse processes is preferred. The used surfactant efficiency has been investigated in the study of Ann. Two runs of soil washing (the first run used fresh surfactant solutions and the second used reused surfactant with the addition of fresh surfactant) were assumed in their study. As the conclusion, the requirement of surfactant could be reduced from 265 g to 86.6 g using 9.1 g activated carbon to achieve 90% removal of PAH with the water/soil ratio of 10 for treat 10 L washing solution [55]. Thus, the total requirement of surfactant could be reduced greatly by AC adsorption.

For economic and environmental reasons, the spent AC must undergo several cycles of regeneration. After PAH reaches breakthrough time, the activated carbon was regenerated using microwave irradiation. The evolution of TX100 and PAH regeneration efficiency as a function of the cycles can be seen in Fig. 7. After five adsorption regeneration cycles, the adsorption capacities for TX100 and PHE were maintained at approximate 418 and 38.2 mg/g, indicating that the microwave irradiation carried out does not significantly affect the surface chemistry of the AC. The carbon regeneration efficiency was maintained at 65.72–90.33% at the sixth cycle and AC cannot be retreated for the adsorption capacity of PHE has reduced 24%. The results indicate that microwave irradiation of activated carbon is a feasible and the cost of activated carbon can reduce 1/5 by regeneration, thereby ensuring its economic and environmental acceptability.

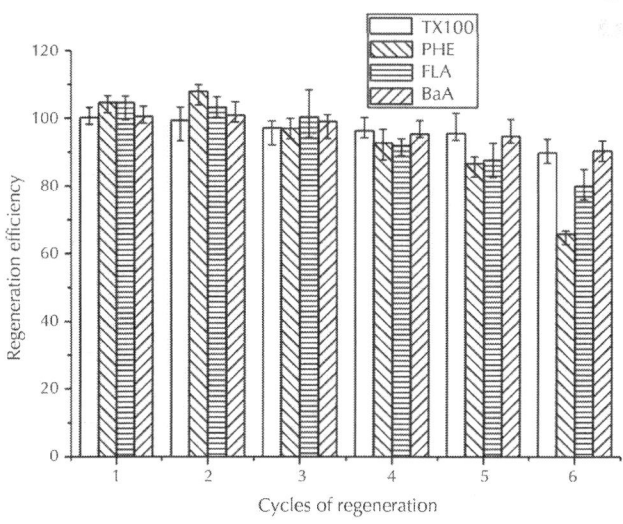

Figure 7: Evolution of the regeneration efficiency of saturated AC.

Costs were estimated based on the currently available guidelines for equipment and operation costs. The operating costs were mainly energy consumption and have minor impact on the overall costs at the experimental scale. Labors, maintenance, and transport costs were excluded from the analysis. The costs difference between whether there are reuse surfactant process or not, mainly depends on the used surfactant efficiency, the costs of surfactant and AC, AC recovery cost, the amount of mass of surfactant and AC. As shown in Table 7, the total costs can reduce $ 0.57–0.58 per 10 L washing solution with AC adsorption. Additionally, more economic benefits can be achieved without the costs of washing solution treatment works. Therefore, adsorption technology can be used as a simple and economic surfactant recovery method in surfactant-enhanced remediation.

Table 7: Comparison of costs under different case

	Washing process		Adsorption process		Regeneration process		Total	Reference
	M_{surf} (g)	Price ($/kg)	M_{carbon}	Price ($/kg)	Cycle	Costs ($)	Costs ($)	
Case 1	265	3.24	–	–	–	–	0.86	[55]
Case 2	86.6	3.24	9.1	0.8	–	–	0.29	[55]
Case 3	86.6	3.24	7.94	0.8	5	<0.005	0.28	This study

M_{surf}: the mass of surfactant; M_{carbon}: the mass of AC.
Case 1: washing without surfactant recovery; Case 2: washing with surfactant recovery by model simulation; Case 3: washing with fixed bed adsorption and regeneration.

CONCLUSIONS

The adsorption behaviors of single and mixed PAH systems in TX100 solution by AC in batch and fixed-bed experiments were investigated. For PAH, experimental adsorption data from single and ternary PAHs in TX100 solution were successfully fitted with the Langmuir isotherm model. This result indicates that an adsorbate monolayer was established during saturation. The adsorption capacity decreased in the mixed PAH system compared with the single PAH system in surfactant solutions as a result of the interactions between components in the mixed system. Given that the estimated E_a was lower than 8, the adsorption of PAHs by AC can be considered as a physical adsorption process. The adsorption of PAHs in surfactant solutions by AC was well described by the pseudo-second-order kinetics model. In TX100 solution, the presence of PAHs did not affect the isotherm shape and the adsorbed amount, indicating that the adsorption value of TX100 was not affected by co-existing PAHs. In the fixed AC column adsorption, the effluent concentration of TX100 can retain over 90% of the initial concentration, and the concentrations of the PAHs were below 10% of their initial concentrations within a longer period. This behavior indicates that the fixed AC column can be used as an alternative method for surfactant recovery in the treatment of water during soil washing. After five adsorption regeneration cycles, the adsorption capacities for TX100 and PHE were maintained at approximate 418 and 38.2 mg/g using microwave irradiation. Based on the rough economic analysis at the experimental scale, the total costs can reduce $ 0.57–0.58 per 10 L washing solution with AC adsorption.

ACKNOWLEDGMENTS

This research was supported by Beijing Natural Science Foundation: mechanisms of selective recovery of surfactant in soil washing solutions with AC (8122027).

REFERENCES

1. A.M.G. Sehili, Lammel Atmospheric Environment 37 (2007) 8301.

2. C.E. Bostrom, P. Gerde, A. Hanberg, B. Jernstrom, C. Johansson, T. Kyrklund, A. Rannug, M. Tornqvist, K. Victorin, R. Westerholm, Environmental Health Perspectives 3 (Suppl.) (2002) 451.
3. G. Shen, W. Wang, Y. Yang, C. Zhu, Y. Min, M. Xue, J. Ding, W. Li, B. Wang, H. Shen, R. Wang, X. Wang, S. Tao, Atmospheric Environment 39 (2010) 5237.
4. N.R. Khalili, P.A. Scheff, T.M. Holsen, Atmospheric Environment 4 (1995) 533.
5. B. Gao, J. Yu, S. Li, X. Ding, Q. He, X. Wang, Atmospheric Environment 39 (2011) 7184.
6. S.D. Liu, X.H. Xia, L.Y. Yang, M.H. Shen, R.M. Liu, Journal of Hazardous Materials 1– 3 (2010) 1085.
7. E. Ferrarese, G. Andreottola, I.A. Oprea, Journal of Hazardous Materials 1 (2008) 128.
8. A.R.U. Johnsen, Karlson Applied Microbiology and Biotechnology 4 (2004) 452.
9. S. Laha, B. Tansel, A. Ussawarujikulchai, Journal of Environmental Management 1 (2009) 95.
10. C.N. Mulligan, R.N. Yong, B.F. Gibbs, Engineering Geology 1–4 (2001) 371.
11. S. Paria, Advances in Colloid and Interface Science 1 (2008) 24.
12. D. Lee, R.D. Cody, D. Kim, Separation and Purification Technology 1 (2002) 77.
13. J.H. Harwell, D.A. Sabatini, R.C. Knox, Colloids and Surfaces A: Physicochemical and Engineering Aspects 1/2 (1999) 255.
14. F.I. Khan, T. Husain, R. Hejazi, Journal of Environmental Management 2 (2004) 95.
15. J.H. Harwell, D.A. Sabatini, C. Chnng, J.H. O'Gaver, T.J. Simpkin, Reuse of Surfactant and Cosolvents for NAPL Remediation, Lewis Publishers, Boca Raton, 1999.
16. H.F. Cheng, D.A. Sabatini, Separation Science and Technology 3 (2007) 453.

17. C.K. Ahn, Y.M. Kim, S.H. Woo, J.M. Park, Chemosphere 11 (2007) 1681.
18. C.K. Ahn, S.H. Woo, J.M. Park, Chemical Engineering Journal 2 (2010) 115.
19. C.K. Ahn, S.H. Woo, J.M. Park, Carbon 11 (2008) 1401.
20. J.Z. Wan, L.N. Chai, X.H. Lu, Y.S. Lin, S.T. Zhang, Journal of Hazardous Materials ½ (2011) 458.
21. D.A. Edwards, R.G. Luthy, Z. Liu, Environmental Science & Technology 1 (1991) 127.
22. J.L. Li, B.H. Chen, Chemical Engineering Science 14 (2002) 2825.
23. W. Zhou, X. Wang, C. Chen, L. Zhu, Colloids and Surfaces A: Physicochemical and Engineering Aspects 0 (2013) 122.
24. M. Abu-Zreig, R.P. Rudra, W.T. Dickinson, L.J. Evans, Journal of Contaminant Hydrology 3/4 (1999) 249.
25. H. Li, Removal Mechanisms of PAHs Contaminants by Soil Washing Remediation, Beijing Normal University, 2010 (Ph.D) 55.
26. L.C. Sander, S.A.W. NIST Special Publication 922.
27. W.E. May, S.P. Wasik, D.H. Freeman, Analytical Chemistry 7 (1978) 997.
28. R.W. Walters, R.G. Luthy, Environmental Science & Technology 6 (1984) 395.
29. X. Liu, X. Quan, L. Bo, S. Chen, Y. Zhao, Carbon 2 (2004) 415.
30. R.J. Robson, E.A. Dennis, the Journal of Physical Chemistry 11 (1977) 1075.
31. P. Levitz, H. Van Damme, D. Keravis, the Journal of Physical Chemistry 11 (1984) 2228.
32. S. Guha, P.R. Jaffe´, C.A. Peters, Environmental Science & Technology 7 (1998) 930.
33. D.P. Prak, H. Pritchard, Water Research 14 (2002) 3463.
34. J.D. Rouse, T. Morita, K. Furukawa, B.J. Shiau, Colloids and

Surfaces A: Physicochemical and Engineering Aspects 3 (2008) 180.
35. I. Langmuir, Journal of the American Chemical Society 38 (1916) 2221.
36. H. Freundlich, Physical Chemistry 57 (1906) 385.
37. M.J.Yuan, S.T. Tong, S.Q. Zhao, C.Q.Jia, Journal of HazardousMaterials 1–3 (2010) 1115.
38. J. Lemic, M. Tomasevic-Canovic, M. Adamovic, D. Kovacevic, S. Milicevic, Microporous and Mesoporous Materials 3 (2007) 317.
39. Y. Al-Degs, M. Khraisheh, S.J. Allen, M.N. Ahmad, G.M. Walker, Chemical Engineering Journal 2/3 (2007) 163.
40. A.R. Kul, H. Koyuncu, Journal of Hazardous Materials 1–3 (2010) 332.
41. A.R. Cestari, E. Vieira, G.S. Vieira, L.E. Almeida, Journal of Colloid and Interface Science 2 (2007) 402.
42. C. Gonzalez-Garcia, M.L. Gonzalez-Martin, V. Gomez-Serrano, J.M. Bruque, L. Labajos-Broncano, Carbon 6 (2001) 849.
43. R. Denoyel, J. Rouquerol, Journal of Colloid and Interface Science 2 (1991) 555.
44. X. Zhu, T. ZhaoGu, Journal of the Chemical Society 11 (1988) 3951.
45. P. Somasundaran, S. Krishnakumar, Colloids and Surfaces A: Physicochemical and Engineering Aspects 0 (1997) 491.
46. N. Erdinc, S. Gokturk, M. Tuncay, Colloid Surface B 1 (2010) 194.
47. P. Punyapalakul, S. Takizawa, Water Research 17 (2006) 3177.
48. S. Lagergren, Vetenskapsakademiens Handlingar 1 (1898).
49. V.P. Vinod, T.S. Anirudhan, Water, Air & Soil Pollution 1 (2003) 193.
50. C. Valderrama, J.L. Cortina, A. Farran, X. Gamisans, C. Lao, Journal of Colloid and Interface Science 1 (2007) 35.

51. J. Yang, K. Baek, T. Kwon, J. Yang, Journal of Industrial and Engineering Chemistry 6 (2009) 777.
52. S. Peng, W. Wu, J. Chen, Chemosphere 8 (2011) 1173.
53. C.K. Ahn, Y.M. Kim, S.H. Woo, J.M. Park, Journal of Hazardous Materials 1–3 (2008) 153.
54. R. Ló pez-Vizcaı́no, C. Sáez, P. Cañizares, M.A. Rodrigo, Separation and Purification Technology 0 (2012) 46.
55. C.K. Ahn, M.W. Lee, D.S. Lee, S.H. Woo, J.M. Park, Journal of Hazardous Materials 1 (2008) 13.

Citations

CHAPTER 1

Jamie Whelan, Ionut Banu, Gisha E Luckachan, Nicoleta Doriana Banu, Samuel Stephen, Anjana Tharale, kshmy, Saleh Al Hashimi, Radu V Vladea, Marios S Katsiotis, and Saeed M Alhassan, Influence of Decomposition Time and H2 Pressure on Properties of Unsupported Ammonium Tetrathiomolybdate-derived Mos2 Catalysts, doi:10.1186/s40543-014-0043-0.

CHAPTER 2

Anne Simon, Sibylle X Maletz, Henner Hollert, Andreas Schäffer, and Hanna M Maes, Effects of Multiwalled Carbon Nanotubes and Triclocarban on Several Eukaryotic Cell Lines: Elucidating Cytotoxicity, Endocrine Disruption, and Reactive Oxygen Species Generation, doi:10.1186/1556-276X-9-396.

CHAPTER 3

M. Ojeda-Morales, M. Hernández-Rivera, J. Martínez-Vázquez, Y. Córdova-Bautista and Y. Hernández-Cardeño, "Optimal Parameters for in Vitro Development of the Fungus Hydrocarbonoclastic Penicillium sp," Advances in Chemical Engineering and Science, Vol. 3 No. 4A, 2013, pp. 19-29. doi: 10.4236/aces.2013.34A1004.

CHAPTER 4

Silva, M. , Carneiro, L. , Silva, J. , dos Santos Oliveira, I. , Filho, H. and de Oliveira Almeida, C. (2014) An Application of the Taguchi Method (Robust Design) to Environmental Engineering: Evaluating Advanced Oxidative Processes in Polyester-Resin Wastewater Treatment. American Journal of Analytical Chemistry, 5, 828-837. doi:10.4236/ajac.2014.513092.

CHAPTER 5

G. Cordeiro, S. Dantas, L. Vasconcelos and R. Brito, "Effect of Two Liquid Phases on the Separation Efficiency of Distillation Columns," Advances in Chemical Engineering and Science, Vol. 3 No. 1, 2013, pp. 1-8. doi:10.4236/aces.2013.31001.

CHAPTER 6

Zhongxiang Chen, Yibin Yan, Said S.E.H. Elnashaie, Catalyst deactivation and engineering control for steam reforming of higher hydrocarbons in a novel membrane reformer, Chemical Engineering

Science, Volume 59, Issue 10, May 2004, Pages 1965-1978, ISSN 0009-2509, http://dx.doi.org/10.1016/j.ces.2004.01.046.

CHAPTER 7

Vito Librando, Matteo Pappalardo, In silico bioremediation of polycyclic aromatic hydrocarbon: A frontier in environmental chemistry, Journal of Molecular Graphics and Modelling, Volume 44, July 2013, Pages 1-8, ISSN 1093-3263, http://dx.doi.org/10.1016/j.jmgm.2013.04.011.

CHAPTER 8

Koyo Norinaga, Olaf Deutschmann, Klaus J. Hüttinger, Analysis of gas phase compounds in chemical vapor deposition of carbon from light hydrocarbons, Carbon, Volume 44, Issue 9, August 2006, Pages 1790-1800, ISSN 0008-6223, http://dx.doi.org/10.1016/j.carbon.2005.12.050.

CHAPTER 9

Jianfei Liu, Jiajun Chen, Lin Jiang, Xue Yin, Adsorption of mixed polycyclic aromatic hydrocarbons in surfactant solutions by activated carbon, Journal of Industrial and Engineering Chemistry, Volume 20, Issue 2, 25 March 2014, Pages 616-623, ISSN 1226-086X, http://dx.doi.org/10.1016/j.jiec.2013.05.024.

Index

A

Activated carbon (AC) 245
Advanced oxidative process (AOP) 101
Advanced Oxidative Processes (AOPs) 100
Ammonium tetrathiomolybdate (ATM) 3
Analysis of variance effects (ANOVA) 112
Anthracene (ANT) 191
Anthraquinone (ANQ) 191

B

Bioremediation 180, 201, 209
Brunauer–Emmett–Teller (BET) 247

C

Carbon fiber reinforced carbon (CFC) 212
Carbon nanotubes (CNT) 26, 27
Chemical oxygen demand (COD) 100
Chemical Oxygen Demand (COD) 109, 111, 119
Chemical vapor deposition (CVD) 212
Chemical vapor infiltration (CVI) 213, 233
Circulating fluidized bed membrane reformer (CFBMR) 141, 144
Colony Forming Unit (CFU) 77
Colony forming units (CFU) 80
Commercial plant 122, 125, 136
Coordinatively unsaturated site (CUS) 17
Critical micelle concentration (CMC) 245

D

Deutsche Forschungsgemeinschaft (DFG) 233
Dibenzothiophene (DBT) 2, 5

Diffuse reflectance infrared Fourier transform (DRIFT) 6
División Académica de Ingeniería y Arquitectura (DAIA) 88

E

Electron-donor-acceptor (EDA) 50
Endocrine-disrupting compounds (EDC) 48
Enzyme Commission (EC) 200
Enzyme-linked immunosorbent assay (ELISA) 35, 37
Equation of State (EOS) 128
Equilibrium between liquid-liquid-vapor phases (ELLV) 128
Estrogen receptors (ER) 48
Experimental units (EU) 77

G

Greenhouse gas 144
Gross National Product (GNP) 71

H

Hydrodesulfurization (HDS) 2, 3
Hydrophobic organic compounds (HOCs) 244
Hydroxyl radicals (HOŸ) 102

I

Ionisation potential (IP) 191

M

Multiwalled carbon nanotubes (MW-CNT) 27

N

Naphthalene dioxygenase (NDO) 193
Natural organic matter (NOM) 29

P

Parts Per Million (ppm) 128

Phosphate-buffered saline (PBS) 34
Plug flow reactor (PFR) 151
Polycyclic aromatic hydrocarbons (PAH) 244
Polycyclic aromatic hydrocarbons (PAHs) 180, 203
Powdered activated carbon (PAC) 101

R

Rainbow trout liver cells (RTL-W1) 26
Reactive oxygen species (ROS) 29
Relative light units (RLU) 36

S

Separation Factor (SF) 134
Sequential Quadratic Programming (SQP) 129
Solid medium Cellulose-Agar (SCA) 78, 79
Solid Medium Cellulose-Agar (SCA) 74
Solvent controls (SC) 37
Soybean peroxidase (SBP) 192
Standard deviation (SD) 37
Surface Area (SA) 11
Surfactant-enhanced aquifer remediation (SEAR) 244

T

Techno Plastic Products (TPP) 32
Temperature-programmed reduction (TPR) 6
Total Petroleum Hydrocarbons (TPH) 70, 73, 75
Total pore volume (TPV) 8, 9
Transmission electron microscopy (TEM) 33

U

Ultraviolet (UV) 102
Universidad Juárez Autónoma de Tabasco (UJAT) 88